ROUTLEDGE LIBRARY EDITIONS:
20TH CENTURY S

Volume 6

A CENTURY OF SCIENCE

A CENTURY OF SCIENCE
1851-1951

Edited by
HERBERT DINGLE

Routledge
Taylor & Francis Group

LONDON AND NEW YORK

First published in 1951

This edition first published in 2014
by Routledge
2 Park Square, Milton Park, Abingdon, Oxfordshire OX14 4RN

and by Routledge
711 Third Avenue, New York, NY 10017

First issued in paperback 2016

Routledge is an imprint of the Taylor & Francis Group, an informa business

British Library Cataloguing in Publication Data
A catalogue record for this book is available from the British Library

ISBN: 978-0-415-73519-3 (Set)
ISBN 13: 978-1-138-96540-9 (pbk) (Volume 6)
ISBN 13: 978-1-138-01353-7 (hbk) (Volume 6)

Publisher's Note
The publisher has gone to great lengths to ensure the quality of this book but points out that some imperfections from the original may be apparent.

Disclaimer
The publisher has made every effort to trace copyright holders and would welcome correspondence from those they have been unable to trace.

A
CENTURY *of* SCIENCE

1851-1951

Written by
SPECIALIST AUTHORS
under the Editorship of
HERBERT DINGLE, D.Sc.

HUTCHINSON'S
SCIENTIFIC AND TECHNICAL PUBLICATIONS
London   New York   Melbourne   Sydney   Cape Town

First Published - February 1951
Reprinted   - - - June 1951

Printed in Great Britain by
William Brendon and Son, Ltd.
The Mayflower Press (late of Plymouth)
at Bushey Mill Lane
Watford, Herts.

CONTENTS and AUTHORS

LIST OF ILLUSTRATIONS

PREFACE

THE Great Exhibition of 1851 was largely a celebration of the achievements of science and its applications. The Festival of Britain of 1951 is not so manifestly a scientific occasion, yet in even greater degree it is permeated by the spirit of science, for this is pre-eminently a scientific age. On the surface, the change that a hundred years have brought about appears anything but pleasant. The attitude of expectation and hope that marked the mid-nineteenth century reaction to the progress of science has given place to one of apprehension, if not of despair. The forces that man seemed more and more to be bringing under his control and ordering to his own desires now turn round and threaten him with the weapons with which he has armed them; the slave has suddenly shown himself as a potential master. We no longer see an endless vista of inevitable advancement stretching out before us; instead we 'dimly guess what time in mists confounds'.

If this situation is to be relieved, one thing at least that must be done is to survey the change that has come about and see what it is. Then and then only can we hope to understand it, and when and only when we have understood it can we hope in reality to achieve the mastery of our own accomplishments that we once thought we had. It is the purpose of this book to assist the first step in this threefold programme. An attempt has been made to recapture the] outlook (i.e. the purely scientific outlook, not the wider generalizations that often resulted from it) of the men of science of 1851, to compare it with that of the present day and to follow the path by which the first has been transformed into the second. The interpretation of the change in terms of ultimate values is left to the reader, but one thing may be noted. In science itself—i.e. our knowledge of the content and interconnectedness of the world of experience—the expectations of 1851 have not been falsified, they have been realized beyond all possibility of conjecture. It is therefore to influences from outside science that we must look for the origin of our fears and misgivings.

The extent, and to some degree the character, of the change in science reveals itself at the very beginning of the task of making the comparison. How shall we subdivide the field to be covered? The classification most suitable for 1851 becomes absurd for 1951; that for 1951 is impossible for 1851. In the earlier years physics fell naturally into the conventional groups—mechanics, heat, light, magnetism and electricity, and so on; these were separate sciences, in very different stages of development, each following its own line of growth independently of the others. Today the boundaries between them have almost completely gone. New unities have arisen—energy, field physics, and the like—almost unheard of before, which spread themselves indiscriminately over the whole array of once separate subjects. Biology tells the same tale. Botany, zoology, entomology can no longer be kept apart; their differences disappear in the vastly greater scope of the conceptions of evolution and heredity. Psychology in 1851 was scarcely a science at all; today it has perhaps the most momentous prospects of all. The general picture is one of extension, unification and generalization. Today there are only three sciences—physical, biological and psychological —that can show dividing lines as sharp as that which in 1851 separated, say, optics and electricity and many other now completely amalgamated departments of research.

In these circumstances it is clear that a detailed comparison is out of the question, and this is further emphasized by the consideration that the mere amount of knowledge now in our possession, apart altogether from its correlation, has, at a modest estimate, far more than doubled itself in the last hundred years. It would therefore have been hopeless to attempt in this book either to produce a complete conspectus of science or, on the other hand, to avoid in many instances presenting the same facts from different points of view. Accordingly no space has been wasted in striving after these impossibilities. The informed reader will have no difficulty in thinking of advances, which he rightly considers important, that are not recorded here, and he will also notice that sometimes the same developments are shown as contributing to two or three different sections of the whole field. Such anomalies are inevitable when one has to compare a little knowledge, divided naturally into many parts, with much knowledge divided only artificially, for purposes of convenience, into a comparable number. The book is to be judged by its success in portraying the broad outlines of a century's progress in science, and not by a conventional and preconceived notion of the way in which it should do so.

In pursuance of the same general aim, the contributors have been given perfect freedom, within the space available, to develop their subjects in their own way. No uniformity has been sought beyond that necessarily imposed by the common objective. To have arranged otherwise would have been to miss an opportunity. The authors are all men eminently qualified to write on their respective fields and most of them have themselves made distinguished contributions to the advances which they have described. They are not men who, however willing they might have been to do so, could conform to a standard pattern of presentation without loss to the reader. The reader, for his part, is thereby sometimes called upon to readjust his attitude as he passes from one chapter to another. It is believed that the effort will be to his advantage.

The Great Exhibition of 1851 was an 'Exhibition of the Works of Industry of all Nations'. There is no human undertaking more thoroughly international in essence and in practice than the pursuit of science, and the Festival of Britain, no less than its predecessor, is a Festival of the planet Earth. This survey of a hundred years' progress strikes the same note. Nevertheless, the reader, if he is British, may well feel a glow of national pride at the share which his countrymen have taken in the whole achievement. Nor is their influence on the wane. If there is to be a similar celebration in 2051 there is every prospect that the British contribution will be no less prominent and vital than in the earlier events.

HERBERT DINGLE

January, 1951

CHAPTER I

THE CONCEPT OF ENERGY

by W. WILSON, D.Sc., PH.D., F.R.S.

IN 1851 the notion of energy was already assuming definite form, not very different indeed, if we regard only familiar phenomena, from the present-day view of it. William Thomson, afterwards Lord Kelvin[1] (1824–1907), introduced the name *energy* (ἐνέργεια) in its present sense about 1854; but the term had been used much earlier in related senses. In 1854 it meant *capacity for doing work* in the sense of the term 'work' explained below, and was indeed almost identified with work. In fact the measure of energy was then and still is work.

A rough and imperfect but not incorrect idea of the way in which energy appeared to Joule, Mayer, William Thomson, von Helmholtz and other scientific worthies of that earlier time is easily conveyed by reference to the common experience of many of us. When we have the roof of our house repaired we have to pay the builder and draw on our stock of capital or wealth. *The work of repairing the roof is not done for nothing.* This stock on which we draw simulates the energy of the physical world. One of its most important characteristics—already known to the natural philosophers of 1851—namely its conservation, is also simulated by our common experience: what we have drawn from our stock of wealth is not lost; but is conserved, as it were, in the enhanced value of the house.

Another characteristic of energy is its tendency to assume a form in which it is *unavailable* for doing work, just as the water which drives a mill, for instance, is no more at the miller's disposal after it has left the mill wheel. The form of energy called heat is

[1]His greatest achievements were accomplished while he was still known as William (or Sir William) Thomson and reference will usually be made to him in this older style.

like this: the lower the temperature the less its availability for doing work. Energy is continually becoming unavailable for doing work. It is still there in undiminished quantity and can still be estimated in terms of work; but cannot actually be used for doing work. William Thomson described this decreasing availability as the degradation of energy.

Origin of the Concept of Energy

The notion of energy and the belief in its conservation grew partly out of futile efforts to make a *perpetuum mobile*. An engine capable of driving the machinery in a factory without needing to be supplied with fuel and without needing outside assistance in any form, or to use up anything that might be within it, would constitute a *perpetuum mobile*. It would in fact be a machine that could do work for nothing, commonly spoken of as a perpetual motion machine.

A simple illustration of the special sense in which the term work is used in physical science is provided by the old-fashioned kind of clock whose mechanism is driven by a descending weight. The work is measured by the product of the weight, expressed perhaps in pounds, and the vertical distance through which it descends, measured perhaps in feet, or in terms of any other convenient unit. When a weight of, say, seven pounds descends through a vertical distance of, say, four feet, the work it does is expressed as, or measured by, 7×4=28 *foot-pounds*. The technical term *force* is used for the weight, and work is always measured by multiplying a force by a distance, or in some equivalent way.

The weight of a body is a very special type of force: it is the downward (gravitational) force exerted on it by the earth and it is important to distinguish it from the mass of the body. When a grocer is 'weighing' coffee or sugar, what he is really doing is measuring the *mass* of the coffee or sugar. These commodities, if carefully preserved, would not lose any of their mass if transported to the equator; but they would certainly have a smaller weight there. A satisfactory definition of mass is too long to be given here; but quite roughly it may be described as a measure of the quantity of matter in a body. Mass, like energy, is conserved— thus we have the law of conservation of mass, more usually called conservation of matter a century ago.

For scientific purposes a unit of force, the *dyne* (δύναμις) has been universally adopted. It may be described as the force which causes the product of the mass (in grammes) and velocity (in centimetres per second) of the body on which it is acting to increase by the unit amount in one second. The units of length, mass and time which have been adopted for scientific purposes are respectively the centimetre, gramme and mean solar second—the second of everyday life. The scientific measure of work is therefore force (expressed in dynes)×distance (expressed in centimetres). The unit, i.e. one dyne × one centimetre, is called the erg (ἔργον).

The successful career of the early steam engine, reinforced by the vast amount of experimentation on heat in the 1840s, also contributed—perhaps more than anything else—to the formation of the concept of energy and to the illumination of obscurities which had in some degree impeded the progress of physical science during the earlier half of the nineteenth century. Heat or caloric, as the old natural philosophers called it, was once thought to be some sort of imponderable fluid and was believed to be conserved—i.e. it was believed that it could neither be created nor destroyed—and William Thomson was still adhering to this view in 1848.

Lastly, the formation of the concept of energy and the belief in its conservation were facilitated by the classical mechanics of Sir Isaac Newton. Indeed energy and its conservation are almost implicit in the old mechanics, which, a century ago, was believed to govern all physical phenomena.

Equivalence of Heat and Work

The fact is that heat is *not* conserved. It is consumed, for example, when a steam engine is working. A definite amount of work is done *at the expense* of a definite quantity of heat, the amount done at the expense of one unit of heat being called the *mechanical equivalent* (sometimes *Joule's equivalent*) of heat. Conversely, heat can be generated in consequence of work done (against friction for example) and the amount of work equivalent to a unit of heat is precisely the same whether it is done at the expense of the heat, or the latter is generated by the work. The credit for establishing these facts is due chiefly to a brewer of Salford, James Prescott Joule (1818–89). He found values for the equivalent ranging from 772 to

778 foot-pounds for the quantity of heat which raises a pound of water 1° F. in temperature, in the neighbourhood of 60° F. The correct value appears to be very near to the higher of these two limits—in terms of our present-day units it is quite close to 41,900,000 ergs of work equivalent to the quantity of heat which raises one gramme of water 1° C. in temperature in the neighbourhood of 15° C.—a quantity known as one *calorie*.

Joule's best known experiment was the famous water-churning one, which he repeated with mercury and other liquids. He had a well-lagged cylindrical vessel (calorimeter) which was filled with water, or other liquid, and a coaxial spindle with paddles attached. When the spindle was caused to rotate heat was generated by the friction between the paddles and the water. The rotation of the spindle was effected by descending weights, the product of each weight and the vertical distance it descended giving the measure of the work it did. The heat generated was of course determined by multiplying the rise in temperature of the water in the calorimeter by the number of pounds of water in it (plus, of course, the quantity of water to which the metal of the calorimeter was equivalent). Joule did many other experiments which cannot be described here. He generated heat by compressing air, by electric currents and in other ways and he dealt with all the possible sources of error, which beset calorimetric measurements, with superb skill. His experimental results led to the inevitable inference of the equivalence of heat and work, and this was supported by his continental contemporaries and by many more recent investigators, including the famous American physicist, H. A. Rowland (1848–1901).

The Principle of Conservation of Energy

The earliest adequate statement of this great principle was made by the famous German physicist, Hermann von Helmholtz (1821–94), in a lecture to the Physical Society of Berlin (23rd of July, 1847) on the *Erhaltung der Kraft*, literally, *conservation of force*. Helmholtz subsumed all kinds of energy under one or other of the two *mechanical* forms: *kinetic energy*, i.e. energy of motion, and *potential energy*, i.e. energy of position or configuration. The weights in the clock, for example, possess potential energy after it has been wound up. He called the sum of these two forms of energy, reckoned for an isolated mechanical system, the *innewohnende Kraft* of the system—the term energy, it will be recalled, had not yet been adopted. He recognized that energy (*Kraft*) existed in many forms which were not *obviously* mechanical; heat energy for example, which was believed to be the kinetic energy, or energy of motion, of the invisible ultimate particles or molecules of the hot body; the energy in light waves, thought to be kinetic energy in the luminiferous medium (ether) and potential energy associated with displacements in it; the energy associated with an electric field, a kind of potential energy; chemical energy and so forth.

Nearly all the physical experimentation of the middle years of the nineteenth century contributed to the conviction that the sum total of the energy, of all forms, in a closed system, as measured by the equivalent work or heat, remained constant. This is the Principle of Conservation of Energy. The special form it assumes when applied to thermal phenomena was called by Rudolph Clausius (1822–88) the first main principle of the mechanical theory of heat, or, as we now say, the First Law of Thermodynamics. It affirms that the increment of the internal energy of a body, or a system, is equal to the sum of the energy conveyed to it in the form of heat plus the work done *on* it.

Thermodynamics

In 1824 the Frenchman, Sadi Carnot (1796–1832), had developed an important theory of heat and of the efficiency of heat engines from premises partly based on the old caloric theory. William Thomson was so impressed by this that he could not accept Joule's interpretation of his experimental results, which of course was in conflict with the conservation of caloric. Joule too considered his results to be incompatible with Carnot's theory; but he proposed that the latter should be abandoned.[1]

Carnot's theory was very simple. It appeared to him that work can only be done by a heat engine (a steam engine for example) when it receives heat from a source at a higher temperature and rejects it to a 'sink' (a condenser, or the atmosphere) at a lower one. He pictured the functioning of the engine as very similar to that of a mill wheel. The mill wheel is driven by the descent of water from

---

[1] *Phil. Mag.*, Vol. XXVI (1845).

a higher to a lower level and the water is not consumed thereby. Just as much water leaves the wheel at the lower level as impinges on it at the higher one. So with the heat engine: heat descends, in driving the engine, from a higher level (temperature) to a lower one, just as does the water in driving the mill wheel, and Carnot assumed that the heat (caloric), like the water, was not consumed, but remained unchanged in quantity—conservation of caloric. "*La production de la puissance motrice est donc due, dans les machines à vapeur, non à une consommation réelle du calorique, mais à son transport d'un corps chaud à un corps froid. . . .*"[1]

He was led, in his reflections about heat engines, to the brilliant conception of a reversible engine, or a reversible cycle of operations, and reversibility is perhaps the most prominent notion in thermo-dynamical treatises, even today. If such a thing as a perfectly reversible mill wheel were possible, one which, if driven backwards would go through *all* its operations exactly in the reverse sense, no other mill wheel working between the same two levels could be *more efficient*, i.e. do more work for the same supply of water; because if there really were a more efficient one, it would be possible to make a *perpetuum mobile* by combining it in a suitable way with the reversible one and the impossibility of a *perpetuum mobile* was one of Carnot's premises. It follows, moreover, that all reversible mill wheels must be equally efficient. Guided by this analogy of the mill wheel, Carnot concluded that all reversible heat engines, working between the same two temperatures, are equally efficient and that no heat engine can be more efficient than a reversible one. This is Carnot's principle, the soundness of which is not questioned today.

Carnot's theory was in 1834 put into admirable form by E. Clapeyron, who also introduced the graphical representation of the cycle of operations (probably suggested by Watt's indicator diagram) of a reversible engine, which still appears even in the most recent text books. There is no doubt that his methods guided both Clausius and Thomson to the correct view of Carnot's principle, though he himself (Clapeyron), it seems, always adhered to the old caloric theory. James Thomson (William's elder brother) deduced from Clapeyron's equation, which expresses the gist of Carnot's theory, almost the exact amount, as subsequent experiment showed, by

[1]S. Carnot: *Réflexions sur la Puissance Motrice du Feu. . . .* (1824).

which the melting point of ice would be lowered by pressure, and it was this especially which convinced William Thomson of the validity of Carnot's principle. This explains why Joule's experiments caused him such perplexity. They indicated, apparently, that caloric was *not* conserved while the principle, the soundness of which could hardly be questioned, seemed to require its conservation.

Both Thomson and Clausius solved the problem independently of one another and almost at the same time (about 1851). In effect they discovered that Carnot's premise of the conservation of caloric was not essential for the deduction of his principle, and each of them succeeded in finding a different and more acceptable foundation for it, and one which was not in conflict with the views of Joule and Mayer about heat. Their arguments cannot be set out here; but it may be said that Clausius's chief premise was that heat cannot pass, of itself, from a lower to a higher temperature; while Thomson's chief premise was perhaps a better one: namely that it is impossible for a self-acting engine to continue indefinitely doing work by cooling a source of heat below the lowest temperature in the surrounding neighbourhood. The impossible type of engine described in this axiom of Thomson's (in the more perfect form which Max Planck, 1858–1947, has given it) is often called the *perpetuum mobile of the second kind* (Wilhelm Ostwald). Carnot's principle is equivalent to the Second Law of Thermodynamics.

One of William Thomson's greatest achievements was to use Clapeyron's equation to *define* a scale of temperature in a way that was independent of the peculiarities of the particular physical quantity or device that might be used in measuring temperature. This he did in two quite different ways. Clapeyron's equation is really an equation expressing the efficiency of a reversible heat engine working between temperatures which are very close together, and it naturally contained an unknown function of the temperature, *Carnot's function*.[1]

Thomson obtained his earlier scale (which, by the way, had its zero at minus infinity) by requiring Carnot's function, C, to be a

---

[1]The equation may be written

$$\frac{\delta p_v . \Delta V}{\Delta Q} = \frac{dt}{C} \; ;$$

C is Carnot's function (though this name is sometimes given to $1/C$).

constant, and his later one, virtually, by identifying it with the temperature. This latter, known as Kelvin's work scale of temperature, is now universally adopted. It happens (fortunately) to be identical with the temperature as it would be given by a perfect gas, if we were to take the product

$$\text{pressure} \times \text{volume}$$

of a fixed quantity of it as a measure of its temperature.

Clausius too had a further great achievement to his credit: he discovered the quantity called *entropy* (1854) which is perhaps as important as energy. When a quantity of heat is supplied to a body, or to a system, in a reversible way and at constant temperature, the quotient of this quantity of heat by the Kelvin or ideal gas temperature is a measure of the consequent increment of the entropy of the body, or system.

The second law may be conveniently stated in terms of entropy as follows: during a reversible change, or changes, the sum total of the entropy of the systems involved remains unchanged; otherwise, i.e. when the change, or process, is *not* reversible, the sum total of the entropy of the systems involved must increase. This increase runs parallel, as it were, with the associated increase in the amount of heat rendered unavailable (at the temperatures at our disposal) for doing work.

The two great principles on which thermodynamics rested exclusively for fifty years were pregnantly stated by Clausius in the following form:

Die Energie der Welt ist constant.

Die Entropie der Welt strebt einem Maximum zu[1].

The two laws of thermodynamics have worked almost miraculously in providing a vast variety of general formulae whose reliability is beyond doubt or cavil, and it is impossible in a small space to give the faintest idea of the wonder of it.

The Statistical Nature of Heat

Gases and vapours are simpler than other forms which materials can assume and they approximate closely to a common form which is called the perfect or ideal gas. Such a gas has comparatively

[1](The energy of the world (universe) is constant. The entropy of the world strives towards a maximum.)

simple properties which suggested long ago that a gas or vapour consists of immense numbers of minute particles (molecules) relatively widely separated from one another on the average, flying about in a random fashion in the containing vessel and exerting almost negligibly small forces on one another. This is the *kinetic theory* which has been abundantly tested and confirmed experimentally since Waterston communicated the first important attempt to construct it to the Royal Society in 1845. Its special interest for us lies in the fact that it is a statistical theory. It can take no account of individual molecules, but only of the assemblage of molecules as a whole and of certain important average values.

One of Waterston's conclusions, not then well founded perhaps, but demonstrated later to be a necessary consequence of classical mechanics, was that the average kinetic energy of translation per molecule is the same for all gases at the same temperature. This is a particular case of the famous principle of equipartition of energy, which emerged from the law of distribution discovered by Clerk Maxwell (1831–79), a Scotsman whose genius rivalled that of Isaac Newton. It was the study of Maxwell's law which revealed to the Austrian physicist Boltzmann (1844–1906) that the entropy of a body, or system, can be identified with the logarithm of the probability of its state; hence the tendency of entropy to increase. A gas, or anything else, will naturally incline to pass from a less probable to a more probable state and in this connection it should be remembered that when large enough numbers are involved probabilities approach certainties.

In 1906 the German physical chemist, Walther Nernst, enunciated a new heat theorem, now frequently termed the Third Law of Thermodynamics. Briefly it gives an absolute value to entropy, whereas previously only entropy changes could be defined or measured. It emerges from the statistical theory if we suppose that there is only one state for a system at the absolute zero of temperature.

It was the statistical view of heat, namely that heat is to be identified with the energy of the microsystems, molecules or what not, of which a body or large system is constituted—to be precise, when statistical equilibrium, i.e. temperature equilibrium, has been established—which led to the first serious departure from the views about energy which prevailed from 1851 till 1900. This came about by the study of black body radiation which, while furnishing a statistical problem with resemblances to that of a gas, brought to light certain facts which could not be reconciled with the principle of equipartition of energy and thus led to a notable departure from the old mechanical ideas of Newton, which had stood for more than 200 years and seemed indeed, fifty years ago, to be established for all time.

Radiant Heat

Radiation is the most important of the processes by which heat is transmitted from one place to another. The heat we get from the sun is radiant heat. Its physical nature is identical with that of light which, till the early years of this century, was regarded as waves in a strange medium called the ether. Radiant energy, or radiant heat, extends from indefinitely large wave-lengths on the one hand to infinitesimal waves on the other. Besides visible light it comprises the long waves used in broadcasting and the very short X-ray and γ-ray waves at the other extreme. The radiation from the sun extends over this indefinitely wide range. In 1859 Gustav Kirchhoff (1824–87), applying the new science of thermodynamics, was able to infer that the character of the radiation inside a vacuous enclosure (cavity radiation), the enclosing wall of which has a constant uniform temperature, *depends only on the temperature*. Its character, he inferred, would be precisely the same as if the enclosing wall were perfectly black, in the sense that it completely absorbed *all* the radiation falling on it. The radiation in such an enclosure is *black body radiation* sometimes called *full radiation* or sometimes just *radiant heat*, and if an opening were made in the wall—small enough to avoid any sensible disturbance of the statistical equilibrium within—the radiation emerging into an exhausted region outside would be just like the radiation proceeding from a perfectly black surface. This device of the cavity and aperture enabled the properties of the radiation to be studied experimentally and Max Planck showed that the observed distribution of energy among different wave-lengths in black body radiation was in conflict with the old classical principle of equipartition of energy. He solved the distribution problem by assuming that the emission and absorption of radiation is a discontinuous process: when radiant heat or energy is emitted it always takes the form of sudden emissions, each one accompanied by a quantity of energy equal to

$$
\left(\begin{array}{c}\text{A certain}\\ \text{universal}\\ \text{constant}\end{array}\right) \times \left(\begin{array}{c}\text{the frequency of}\\ \text{vibration in the}\\ \text{emitted radiation}\end{array}\right)
$$

$$h \qquad\qquad\qquad \nu$$

He represented the universal constant (Planck's constant) by the letter h and named it the quantum of action—hence his theory is called the quantum theory. He found the value of his constant to be about

$$6.5 \times 10^{-27} \text{ ergs} \times \text{seconds.}$$

It is so small that it is of no consequence in macrophysics, though exceedingly important in atomic physics.

A. Einstein (b. 1879) actually suggested that light or radiation *consists of small bundles of energy*, almost like small particles—they are now called *photons*—and forty years ago some phenomena seemed to make it certain that light really had this corpuscular character, while others seemed to make it equally certain that the undulatory view of light was the correct one. The character of black body radiation, it was found, could be accounted for on either view. It can be regarded as a gas whose molecules are photons (both the de Broglies, Bose and Einstein) or alternatively as a wave phenomenon.

The solution of this enigma emerged from the wave mechanics (one of the two equivalent forms of quantum mechanics) of Prince Louis de Broglie (*c.* 1924) which was further developed with great success by Erwin Schrödinger. De Broglie actually predicted that waves would be found to be associated with *all* elementary particles and with electrons in particular, and he also predicted exactly how the associated wave-lengths depended on the mass and velocity of the particles. His predictions were completely confirmed by the experimental work of Sir George Thomson, the Americans Davisson and Germer, and others.

It appears now that the wave phenomena are expressions of

probabilities, the photons, electrons, etc. being the physical entities, and it may be added that the product of the range of frequencies in a wave group (any kind of wave) and the time it takes to travel through its length is of the order of unity. It is a consequence of this that the product of the uncertainty in the measured value of the energy of an electron, for example, and the uncertainty about the exact instant when it has this energy, must be of the order of h, at least. This is a special case of the uncertainty relation of Werner Heisenberg. (See Chapter XX.)

Mass and Energy in the Theory of Relativity

An important consequence of the theory of relativity, abundantly confirmed by observation and experiment, is that the energy of a body, or system, is proportional to its mass. In fact

$$\left\{\begin{matrix}\text{Mass in}\\ \text{grammes}\end{matrix}\right\} \begin{matrix}\text{multiplied twice}\\ \text{in succession by}\end{matrix} \left\{\begin{matrix}\text{Velocity of}\\ \text{light in free}\\ \text{space in cm./sec.}\end{matrix}\right\} = \left\{\begin{matrix}\text{Energy}\\ \text{in ergs}\end{matrix}\right\}$$

$$m \qquad \times \qquad c^2 \qquad = \qquad E$$

Indeed there is an energy equivalent of mass, and if we were to choose such units as would make the velocity of light equal to unity we should be able to write

$$\text{mass} = \text{energy.}$$

In fact the distinction between them, which once seemed so clear and obvious, turns out to be unreal. They are identical. In consequence of this law of Einstein the energy of a system, like its entropy and its mass (which is the same thing as its energy), is now regarded as having an absolute value; whereas in the 1850s and much later only *energy changes* could be defined and measured. The old laws of conservation of mass (or matter) and conservation of energy, still regarded as distinct forty-five years ago, are now believed to express one and the same thing. (*Vide* paragraphs on atomic and nuclear energy in Chapter IV.)

The fact that the sun appears to change so slowly in its surface temperature and in other ways, notwithstanding the enormous rate at which it is losing heat, receives its explanation from this Einsteinian mass–energy law. It is believed to be due largely to the formation of helium from hydrogen. The mass of the helium produced is a trifle less than that of the hydrogen from which it is formed, the difference being slightly more than $\frac{7}{1000}$ gramme for each gramme of helium that is formed. This difference takes the form of radiation and $\frac{7}{1000}$ gramme is equivalent to 250,000 horse power for an hour. The total rate at which the sun is radiating energy is estimated to be equivalent to a diminution in its mass of about 250 million tons every minute; but its mass is so great that even this rate of diminution is not serious. It could last at this rate between two and three thousand billions of years, i.e. 2 or 3×10^{15}, years.

When we remember that all the radiating stars in the universe are throwing away their mass at an average rate perhaps equal to or greater than that of the sun we can appreciate that energy has now acquired a cosmical significance undreamt of a hundred years ago.

CHAPTER II

FIELD PHYSICS
(*Gravitation and Electromagnetism*)

by J. L. SYNGE, M.A., SC.D., F.R.S.

THE elementary phenomena of light and gravitation are familiar
to us all, and have been familiar to mankind from the beginning.
If your legs fail, you fall to the ground, and you shade your eyes
with your hand against the light of the sun.

Electricity and magnetism, on the other hand, are historically
much more esoteric. At the beginning of the nineteenth century
they were (except for the thunderstorm and the mariner's
compass) in the nature of scientific curiosities, and they were
separate and distinct—electricity on the one hand, magnetism on
the other.

Then, in a flood of research in the period 1820–30, Oersted,
Ampère and Faraday showed that electricity and magnetism were
essentially one thing—that an electric current produced a magnetic
field, and that the motion of a magnet caused a current to flow in
a coil of wire. The experimentation was very simple, and out of it
there ultimately came the practical utilization of electricity as a
source of light and power, so familiar to us today.

In 1850, then, at the beginning of the century under review,
there were three things, all utterly distinct from one another and
thought of as belonging to different parts of physics altogether:
 (i) Light.
 (ii) Gravitation.
 (iii) Electromagnetism.

Light in the aether-jelly
By the year 1850 physicists had come to know a great deal about
the way in which light behaves and they had built for themselves
a model in terms of which they could think. The corpuscular theory

24

had been rejected as unsuitable, and light was regarded as a vibration.

The vibration most familiar to mankind is the vibration of the string of a musical instrument; this vibrating source of energy transmits sound waves through the air, causing the sensation of hearing. Sound and hearing are natural analogues of light and sight. We think of a glowing body as a vibrating violin string, the light as sound waves in the air, and the eye as the organ which takes over the function of the ear, seeing instead of hearing.

But the analogy fails. Sound is transmitted only through material media; light crosses a vacuum with the greatest ease. In sound, the air oscillates along the line of transmission; the polarization of light indicates a transverse vibration. Reluctant to abandon a mechanical explanation, men said: "Light is a transverse vibration in an *aether*, which is a sort of incompressible elastic jelly, filling all space, but offering no resistance to the passage of material bodies." The idea of the aether saturated the physics of 1850–1900. On the scientific stage it was the main actor—intangible, unweighable, unknowable, but omnipresent.

Yet those who had grown up with the aether did not regard it as essentially mysterious. Speaking in 1884, Lord Kelvin (1824–1907) compared it to Burgundy pitch or Scottish shoemaker's wax, which can be formed into the shape of a tuning fork and made to vibrate, but which also permits bodies to fall slowly through it. He said: ". . . we need not fully despair of understanding the property of the luminiferous ether.[1] It is no greater mystery at all events than the shoemaker's wax or Burgundy pitch. That is a mystery, as all matter is; the luminiferous ether is no greater mystery."

G. F. FitzGerald (1851–1901), speaking in 1888 as President of the Mathematical and Physical Section of the British Association, had a shorter way with disbelievers:

"People who think a little but not much sometimes ask me, 'Why do you believe in the ether? What's the good of it?' I ask them, 'What becomes of light for the eight minutes after it has left the Sun and before it reaches the Earth?' When they consider that they observe how necessary the ether is."

[1]This spelling was much used as well as *aether*.

Gravitation

Both Kelvin and FitzGerald were on the defensive, for the type of physical explanation they favoured for the propagation of light cut a sorry figure beside the austere simplicity of Newton's theory of gravitation.

Gravity is so familiar to us that it is commonplace, and our primitive wonder is awakened only by the refusal of tight-rope walkers and spinning-tops to fall down. But a visitor from inter-stellar space (where gravitational fields are insignificant) would sit fascinated for hours watching you drop pennies on the floor. He, like all of us, would know from his nursery days that you can push and pull, be pushed and be pulled, but that pushes and pulls can be transmitted only by material contact. He would shake his head disbelievingly at the dropping pennies: "I'm sure it's done with wires," he would say, and grope round for a fine wire pulling the penny down to the floor.

Newton's idea of gravitational attraction without material connection is not in agreement with the push-pull concepts of the nursery. It is something utterly unlike the light-theory of an aether-jelly. And in these two we find embattled two undying enemies, action-at-a-distance (gravity) and action-through-a-medium (light).

Newton's theory of gravitation passed serenely through the storms of the late nineteenth century in splendid isolation, dis-connected completely from light and electromagnetism. No man learned how to switch off gravity with a button or to alter by a tittle the universal attraction of one body for another. All that was done was to continue building the great mathematical structure reared on Newton's Laws of Motion, and his absurdly simple, but weirdly accurate, Law of the Inverse Square.

Light is electromagnetic

In this attempt to outline the progress of field physics for the century 1850–1950, two names stand out: James Clerk Maxwell (1831–79) and Albert Einstein (b. 1879)—a strange coincidence in dates, like Galileo (1564–1642) and Newton (1642–1727).

Maxwell came from Dumfriesshire and was educated at Edin-burgh and Cambridge. He was at one time Professor of Natural Philosophy at King's College, London, and later Professor of

Experimental Physics at Cambridge, where he was the inspiring force behind the establishment of the Cavendish Laboratory.

Michael Faraday (1791–1867) believed strongly in action-through-a-medium as opposed to action-at-a-distance as the proper explanation for the phenomena of electromagnetism. He had great experimental skill and a rich intuitive imagination, but was entirely without skill in mathematical manipulations. Maxwell clothed Faraday's ideas in mathematical form. This does not mean that Maxwell merely carried out a routine translation into mathematics; the theory which Maxwell published in 1873 in his 'Treatise on Electricity and Magnetism' must be regarded as a feat of genius, based on, but far transcending in generality, the ideas of Faraday.

The culmination of Maxwell's theory is a set of partial differential equations, the general laws which all electromagnetic fields obey. They enjoy a unique distinction; begotten long before the theory of relativity, they are unchanged by it. In technical electromagnetism they are the ultimate court of appeal, and recent developments in wireless (particularly in radar) have widened the circle of those acquainted with 'Maxwell's equations'.

But Maxwell's achievement was simpler and more spectacular than that. *He identified light with electromagnetism,* in the sense that he recognized light as an electromagnetic phenomenon. A guess of great daring, based on the near-equality of two numbers.

This was an outstanding discovery. In his 'Treatise', Maxwell set out very clearly and calmly his reason for making this extraordinary identification of things apparently so different. He said, in effect, that light needs an aether and electromagnetism needs an aether, and that a test should show whether those two aethers might not really be one and the same. So he compared the known velocity of light with the velocity that electromagnetic waves should have according to his theory. They were nearly the same (about 186,000 miles per second), and on the basis of that near-coincidence Maxwell concluded that light is an electromagnetic vibration.

This, in his view, did not abolish the aether; on the contrary it made it all the more important. Now the aether is gone, but Maxwell's equations, and his identification of light with electromagnetism, remain.

Light and Radio Waves

Maxwell died without knowing how right he was, for it was not until 1888 that Heinrich Hertz (1857–94) succeeded in producing electromagnetic waves with a wave-length short enough for experimental purposes (about two metres). Hertz found that these waves had the same velocity as that of light and that they interfered with one another like light waves.

The bigger the discovery, the longer it takes to be absorbed into the body of science. Maxwell's discovery was one of the first order, and it needed more than the experiments of Hertz to lead the older generation into new ways. Lord Kelvin seems never to have accepted Maxwell's theory, and it took many years before text-books in optics placed the subject squarely on an electromagnetic basis.

The absorption of new ideas into public consciousness naturally takes longer than their absorption into science, and it is perhaps too soon to expect the world at large to recognize radio waves and sunlight as brothers and to regard the distinction between different wave-lengths in the radio programme as a distinction in colour. Just as red light would change to blue if you were to move fast enough towards its source, so would long-wave radio turn to short-wave if you could move your receiving set fast enough towards the transmitting station. And if you moved faster still towards the transmitting station, you might dispense with your receiving set, because the radio waves would ultimately become visible light! But of course all the velocities involved are fantastically high, so that the 'experiments' described above are in the nature of *jeux d'esprit*. That does not alter the fact that the mathematics of radio waves and the mathematics of light waves are one and the same, with a mere change of one constant—the wave-length.

Days of Confusion

To think that the wedding of light and electromagnetism solved all problems would be quite wrong. On the contrary, it was the beginning of what might be called the 'Days of Confusion'.

Maxwell did not destroy the aether, but strengthened it. His electromagnetic theory of light was based on the identification of two velocities of propagation *through the aether*—the velocity of light and the velocity of electromagnetic waves.

Is the aether at rest? We are now, in 1951, so thoroughly in-doctrinated with the spirit of relativity that, at this question, the youngest schoolboy raises his hand and asks, "Please, Sir, relative to what?" But not so Newton, nor the average scientist of the pre-Einstein era. To them absolute rest had a meaning, and it was completely natural to identify the aether with the frame of reference which is at rest in the absolute sense.

In that case, the earth is ploughing through the aether, changing its velocity as it pursues its orbit round the sun. This fits in perfectly with the observed aberrations of stars—they shift their apparent positions slightly and the telescope, carried by the earth, must be given slightly different directions at different seasons of the year if it is to catch the light rays coming from a star, a phenomenon discovered by Bradley as long ago as 1727.

The observed aberrations indicated that the earth did not carry the aether with it. It should therefore be possible to measure the velocity of the earth through the aether by comparing the velocity of light in different directions, and this was attempted in 1887 in the now-classical experiment of A. A. Michelson (1852–1931) and E. W. Morley (1838–1923). The result was null—the velocity of light was found to be the same in all directions. In aether-language this meant that the earth *was* dragging the aether along with it, although Bradley's aberration meant that it *was not*.

FitzGerald and H. A. Lorentz (1853–1928), in 1893 and 1895 respectively, explained the null-result of the Michelson–Morley experiment by the hypothesis that a moving body is automatically contracted in the direction of the motion, this contraction affecting the Michelson–Morley apparatus and so leading to equal measured velocities in all directions. It was only with the advent of relativity ten years later that the meaning of the FitzGerald-Lorentz contraction became clear. Against the background of an aether-jelly, such an extraordinary hypothesis, created *ad hoc* (though not without an explanation in terms of the current electromagnetic theory) to explain a certain result, could only add to the general confusion of thought.

For nearly a quarter of a century the physics of the aether (luminiferous and electromagnetic) was dogged by this contra-diction—the aether was at the same time dragged by the earth and not dragged by the earth. On this feverish period we look back with the calm of a recovered patient; by what superior wisdom do we make sense out of a contradiction?

The truth of the matter, as we now understand it, is this. To say, "The aether is dragged with the earth", or to say, "The aether is not dragged with the earth", has as little meaning as to say "Fi-fo-fum"; and each of these statements is much more dangerous than "Fi-fo-fum" because it *seems* to mean something. You can drag, or not drag, a sack of potatoes, but you cannot drag, or not drag, the aether, the essential difference being that the potatoes have in their skins, their eyes, their cells, their molecules, their atoms, parts which can be identified now and five minutes hence, whereas the aether has no such identifiable parts and never should have been thought of as having them.

You can follow the motion of anything capable of continual identification, but try, for example, to follow the motion of a wave in a stormy sea. At one moment it seems to have an identity, but it sinks down or becomes merged with other waves, and, although the particles of the water are continuously identifiable (you could mark them with dye), the wave (which is a transient property of the water) has no definite continuing identity, except of course under favourable circumstances. You can ask with some meaning, "Where are now the atoms of oxygen and nitrogen and carbon that issued from the mouth of Julius Caesar?" but you cannot ask with any meaning at all, "Where are now the sound waves that conveyed his commands?"

In the period 1850–1900 and for a few years more, physics was glued to an archaic idea from which it could not unstick itself— the idea that you could carve your name on the aether, or dye it, or in some way recognize again a particle of it, and so follow it and watch it move. Anyone who now thinks relativity abstruse or difficult to understand should ponder over the abstruseness and difficulty of attaching any meaning to 'the motion of the aether', just five empty words.

The New Deal

When Einstein produced the special theory of relativity in 1905, the cure, to some, seemed worse than the disease. The word 'relativity' has spread an intoxication so far beyond the ordinary boundaries of physics that it is interesting to note that the word did not appear in the title of Einstein's paper, which read: "On the Electrodynamics of Moving Bodies". The word 'relativity' came in thus: ". . . the same laws of electrodynamics and optics will be valid for all frames of reference for which the equations of mechanics hold good. We raise this conjecture (the purport of which will hereafter be called the 'Principle of Relativity') to the status of a postulate, and also introduce another postulate, which is only *apparently* irreconcilable with the former, namely, that light is always propagated with a definite velocity c which is independent of the state of motion of the emitting body."

Einstein wiped out the aether in one sentence. "The introduction of a 'luminiferous ether' will prove to be superfluous inasmuch as the view here to be developed will not require an 'absolutely stationary space' provided with special properties, nor assign a velocity-vector to a point of the empty space in which electromagnetic processes take place."

But the kernel of Einstein's paper is not to be found in the title, nor in the word 'relativity' (so vague a word might mean anything), nor yet in the elimination of the aether. Had his paper been called 'Simultaneity—absolute or relative?' it would have brought out the real revolution in thought. For Einstein attacked (and, in the view of science, demolished) the Newtonian concept of an absolute time, and with it the Newtonian idea that the word 'simultaneous' has an absolute meaning. Two events occurring at different places may be simultaneous as judged by you, but not simultaneous as judged by me, and this idea is simple and natural and reasonable except for anyone who has Newton in his bones. In 1905 the word 'now' took on a new meaning, not, it is true, affecting the daily lives of men, but easing physics of the awful burden of confusion it was labouring under.

Einstein avoided the question of 'aether-drag versus no aether-drag' by calmly saying that he would not assign any velocity to a point of empty space. Why should he?

As far as electromagnetism was concerned, the controversy of 'action-at-a-distance versus action-through-a-medium' was settled in an extraordinary and unexpected way. Einstein's relativity might be said to favour action-through-a-medium, since it accepted Maxwell's partial differential equations, but the medium was deprived of its most essential property—that of being at rest or in motion. The robust body of the Cheshire Cat was gone, leaving only a sort of mathematical grin. A new word was needed to describe this emasculated form of action-through-a-medium, and the expression 'field theory' has come into use to describe a theory in which physical effects are transmitted continuously (according to partial differential equations in the language of mathematics), but in which the 'medium' is usually not materialized in the sense that we can speak of its 'velocity'.

Space-time

Naturally this made it harder to think intuitively, and one cannot advance in science without intuitive thought. We might say that in 1905 Einstein shot the horse from under physics, and for three years it had to stumble along on foot, regretting the former days when it had ridden the aether. But in 1908 H. Minkowski (1864–1909) solved the transportation problem for physical thought by providing a bicycle—the geometry of four-dimensional space-time. In a paper showing sketches of space-time, he wrote: "Henceforth space by itself, and time by itself, are doomed to fade away into mere shadows, and only a kind of union of the two will preserve an independent reality."

One senses in Einstein and Minkowski two giants of thought, but thinking differently, Einstein the physicist approaching his goals through strange flashes of intuition very difficult for others to understand, and Minkowski the mathematician polishing, simplifying and extending what had seemed so difficult and involved.

Minkowski gave Einstein full credit for destroying the Newtonian concept of absolute time, but thought the expression 'relativity-postulate' a very feeble one; he preferred the 'postulate of the absolute world', or, briefly, 'world postulate'.

A Geometer's World

In 1854 G. F. B. Riemann (1826–66) laid the basis for the discussion of the geometry of a space of any number of dimensions, and from that time any mathematician and any physicist might have talked over their port of the possibility of thinking of the world as a four-dimensional continuum, with three dimensions of space and one of time. But it would not have got beyond this after-dinner stage, because the physicist would have regarded it as waste of time, seeing in the idea no inspiration for the understanding of nature. Newton had sliced the four-dimensional block of events into a pack of three-dimensional cards (each representing the world as it is at a moment), and it would only be flying in the face of common sense to glue them together again into a single block.

But after 1905 it was different. The Newtonian slicing was disallowed, absolute time was rejected, and you sliced the block into your own pack of cards, according to your frame of reference.

If each man had his own separate world, it would be natural for him to slice it to suit himself, using his own definition of time. But there is only one physical world for all, and although each observer may slice it according to his own plan, there is going to be confusion when two or more observers meet and discuss what they observe in nature. Far better leave the world-block unsliced— that is what Minkowski said in effect—and think of it as a whole in the language of four-dimensional geometry. Einstein's rejection of absolute simultaneity and his assumption of an absolute light-velocity transformed the idle idea of a four-dimensional space-time into a powerful vehicle of thought, but it was Minkowski, not Einstein, who saw this. It is a curious fact that the idea of space-time, so often thought of as roughly equivalent to relativity, was not conceived first by the originator of relativity.

The old aether had been created in men's minds as a vehicle for thought—it was the absolute space of Newton endowed with vibratory properties. The alternative offered by Minkowski was an absolute space-time of four dimensions, an immutable absolute in which all events are carved, and in which the annihilation of a particle is no more than the ending of a world-line, just as a chalk line on a black-board may end. But our minds have been schooled for centuries in the tradition of absolute space and absolute time, and the idea of four-dimensional space-time is still a difficult and abstract idea for practical scientists and for the public generally; it is likely to remain so unless we develop techniques for imparting to children at a tender

C

age velocities comparable with the velocity of light (which Heaven forbid !).

But, in the realm of theoretical physics, where the more practical scientists and the public are not coerced to participate, it is a four-dimensional space-time we are living in—a geometer's world.

Curved Space-Time

By abolishing from science the concept of absolute simultaneity, Einstein cleared up the confusions which had choked the theories of electromagnetism and light-propagation (the same thing, since light is electromagnetic). But he fell foul of Newtonian gravitation, for the law of the inverse square postulates an absolute 'now' for two gravitating bodies, and without absolute simultaneity the law becomes meaningless.

Efforts were made by H. Poincaré (1854–1912) and by Minkowski to adjust the statement of Newton's law of attraction so as to fit in with the relative character of time. But their results were not satisfactory, and the anomaly of having an absolute time in gravitation and a relative time in electromagnetism was not resolved until Einstein in 1916 produced a radically new theory.

On this occasion Einstein did not give the word 'relativity' a secondary position, for he called his paper 'The Foundation of the General Theory of Relativity'. What the paper actually was might have been expressed by some such title as 'Gravitation regarded as Curvature of Space-Time', for gravitation was what the paper dealt with, and the 'curvature' of space-time (as opposed to the 'flatness' of Minkowski's space-time) was the way in which, according to Einstein, a gravitational field manifested itself.

If the concept of four-dimensional space-time is too abstract for public appreciation, the *curvature* of that space-time may seem to take us altogether too far into abstraction. But, without going into details, we may hope to get some outline of what the problem was and how Einstein succeeded in dealing with it.

It must be clearly understood that Newton's theory of gravitation was not a poor theory in need of repair. Discrepancies between theoretical prediction and astronomical observation were few and very small, and no one was seriously worried about them. But accurate as it might be from a practical standpoint, the Newtonian theory of gravitation cut directly across Einstein's basic idea of the equivalence of different frames of reference, each with its own time-system. The objection was ideological, rather than practical.

The more one thinks of the general theory of relativity, the more amazing it becomes. Maxwell stood on Faraday's shoulders. Einstein drew from Minkowski the idea of space-time,[1] and from the Italian mathematicians M. M. G. Ricci (1853–1925) and T. Levi-Civita (1873–1941) their mathematical shorthand, the tensor calculus, without which the work would have been quite unmanageable, but the basic idea he owed to no one but himself.

We can describe the position of a point on a table-top by *two* measurements—the distances of the point from one side and one end of the table; and we can describe the position of a point on the surface of a ball by *two* measurements—latitude and longitude, as on the earth's surface. We say, then, that the table-top and the surface of the ball are both *two*-dimensional. Space-time is *four*-dimensional because we describe an event by *four* measurements, three for position and one for time.

A table-top is flat and the surface of a ball is curved. The ideas of flatness and curvature are not necessarily confined to planes and surfaces, and we can speak (with mathematical meaning) of a *flat* four-dimensional space-time and a *curved* four-dimensional space-time. In this sense, the space-time of Minkowski is flat, like the table-top.

Now in his work in 1905 on the theory of relativity (later called the *special* theory, to distinguish it from the *general* theory of 1916) Einstein had excluded gravitation, thinking only of electromagnetic phenomena, and so it might be said that *flat* space-time corresponds to the *absence* of a gravitational field. What then so natural as to think that possibly a *curved* space-time corresponds to the *presence* of a gravitational field? Such thoughts are easy after the event; before it, they are likely to vanish into thin air, for there is no logic in them, only speculation.

But Einstein carried the idea through, guided by the principle that all co-ordinate systems are equivalent for the description of

[1]There is scant reference to Minkowski in the paper cited, but elsewhere Einstein wrote of ". . . the important idea contributed by Minkowski. Without it the general theory of relativity . . . would perhaps have got no farther than its long clothes."

natural laws. The final theory was, in a certain sense, a generalization of Newtonian gravitational theory, if we consider that theory in the form which Laplace had given it about the year 1780. Where Laplace had *one* quantity (gravitational potential) satisfying *one* equation, Einstein had *ten* quantities (gravitational potentials) satisfying *ten* equations. For the paths of planets and light rays in space-time Einstein took the simplest curves available, namely, those which correspond to straight lines in a plane.

The first danger one would look for would be a wide divergence from Newtonian theory, and the cautious revolutionary might have tried to include in his theory a number of unspecified constants, to which values might afterwards be assigned in order to make the theory fit the facts. Einstein's theory carried only two constants, and he did not trim these to fit but equated them to two fundamental constants of nature—the velocity of light and the constant of gravitation. Without any artificial adjustment at all, the results of the theory were found to reduce very nearly to Newtonian results under the conditions which obtain in the solar system, where the gravitational field is weak (in a technical sense) and the relative velocities of the bodies are small compared with the velocity of light.

The Test of Observation

If that had been all, it would have been enough. The galling inconsistency between absolute time and relative time would have been removed, and the observed astronomical motions would have been predicted as well by the new theory as by the old. But Einstein's theory gave results differing slightly from the Newtonian predictions, and so the case of 'Einstein versus Newton' could be tried on the basis of accuracy of prediction rather than on the basis of ideology (absolute time against relative time) alone.

These differences concern three things—the orbits of the planets, the paths of light rays, and the reddening of the light from an atom situated in a gravitational field. They have been tested for the orbit of the planet Mercury, for a light ray passing close to the sun in eclipse, and for the light received from the sun and from the star called the 'Companion of Sirius'; in all cases the verdict has gone in favour of Einstein, although (on account of the minuteness of the effects) the jury of astronomers has not had an easy task.

Thus Einstein's theory of gravitation improves on Newton's both on the philosophical level (by eliminating absolute time) and on the practical level (by more perfectly predicting natural phenomena).

Later Work

In 1917 Einstein passed beyond the solar system to consider the universe as a whole, modifying his field equations so as to obtain a world of finite size. In doing so, he opened up a domain of research (cosmology, including the theory of the expanding universe) to which many have contributed, including W. de Sitter (1872–1934), G. Lemaître (b. 1894) and H. P. Robertson (b. 1903). The physical observation with which these theories are linked is the reddening of the light received from distant nebulae.

Space remains only to mention the unified field theories which have been sought during the past thirty years by a number of physicists, including notably H. Weyl (b. 1885), Sir Arthur Eddington (1882–1944), E. Schrödinger (b. 1887), and Einstein himself. The idea behind this work is to create a theory in which gravitation and electromagnetism are so intimately mixed that they appear as two different parts of one single concept, but no theory so far developed has won the acclaim accorded to Einstein's earlier work. Behind this somewhat vague aesthetic goal there lies the more concrete one of explaining the magnetic fields possessed by massive rotating bodies such as the earth.

In all this work the thought-pattern is that of geometry, generalized far beyond the simple space-time of Minkowski. For some minds the geometry of four (or more) dimensions is stimulating and satisfactory, but others would like to go back to the material models of the nineteenth century. At present there seems to be no other alternative. Perhaps, lying in his cot or still unborn, there is someone who will find the geometrical way of thinking repellent and will substitute another of which we now know nothing. An article like this, written in 2050, will have his tale to tell, and, if history runs true to form, it will not be a tale of minor revisions, but of many pages torn up and thrown in the fire. And perhaps, out of the pages two will survive—one bearing Maxwell's partial differential equations of the electromagnetic field in vacuo and the other Einstein's partial differential equations of the gravitational field in vacuo. For Lord Kelvin was right—we know more about a vacuum than we know about anything else.

CHAPTER III

PARTICLE PHYSICS

by H. T. FLINT, PH.D., D.SC.

ALTHOUGH matter on a large scale gives to the senses the impression that it has a continuous structure, theories that matter can be divided and sub-divided indefinitely have not met with success. From early times the alternative view has been held that matter consists of a collection of indivisible particles which, being extremely small, give the appearance of continuity on a large scale. This is the view which forms the foundations of the theories of chemistry and physics today, but until a little more than a hundred years ago this theory of matter was on the whole only qualitative.

Then, in the early days of the nineteenth century, the work of Dalton laid the foundations of a quantitative atomic theory. During these years he published his view that the atoms of a particular element were identical in all respects but differed from those of other elements. He considered that chemical combinations occurred in simple numbers of atoms—one to one or one to two, and so on—and this was supported by the quantitative laws of chemistry which were just then being experimentally established.

The fact that some atomic weights are very nearly integral multiples of that of hydrogen led Prout in 1815 to suggest that the atom of hydrogen was the fundamental particle from which all elements were constructed. Nothing smaller than this atom was at that time contemplated. There are some very marked deviations from this integral relation amongst the chemical elements and Prout's hypothesis was for a long time left aside. More recent discoveries and developments have led to a correct interpretation of it, and its significance as an essential law in the theory of matter is now understood.

The atomic theory can be regarded as part of the development of the molecular hypothesis of matter which has a very long history.

Like the atomic theory it was not until the nineteenth century that it became established as a quantitative theory, although it should be recalled that Daniel Bernoulli, in 1738, showed that if fluid pressure on the walls of a container could be regarded as due to the impact of molecules then, provided that the density is not too great, the pressure is proportional to the density. This is Boyle's law for gases.

It is an interesting coincidence that exactly one hundred years ago, in 1851, Joule published a paper in the *Memoirs of the Manchester Literary and Philosophical Society* entitled 'Remarks on heat and the constitution of elastic fluids', in which he made the first calculation of a molecular magnitude. This was the average speed of a gas molecule. But it was not until ten years later that the molecular hypothesis was to reap its greatest success in the theory of gases due to Maxwell and Boltzmann. This is one of the greatest triumphs of classical physics and illustrates the aim and method of physical theory at this period. A physical problem was then considered to be solved when, if the process under consideration could not actually be followed in all its detail, a model could be made or conceived to which known mathematical operations could be applied and results deduced in agreement with experiment.

Kinetic Theory of Gases

An outstanding example of the solution of a problem in this sense is afforded by the kinetic theory of gases, where from a model consisting of a collection of molecules in motion subject to the laws of mechanics, the important quantitative relations concerning pressure, temperature, thermal conductivity and so on were deduced. Thus laws proved to apply to large scale problems were shown to apply to those on the small scale of molecules.

The discovery of the Brownian movement of small particles suspended in liquids or in air, studied quantitatively, especially by Perrin, has revealed the reality of molecular motion, for the suspended particles differ from molecules only in size. They are small enough to show the fluctuations resulting from bombardment by the molecules but large enough to be seen in a microscope.

The beginning of the nineteenth century brought the discovery of the phenomenon of electrolysis, the laws of which were stated by

Faraday about 1840. At this time electricity was thought to be something distinct from matter, but an important contribution which is relevant to the question of the atomicity of matter was the assumption that each univalent atom is associated with a definite charge and a multivalent atom with a multiple of it. This indicates a quantitative relation between electricity and matter.

Cathode rays and Electrons

In the 1870s great interest developed in the cathode ray phenomena of the discharge tube. Cathode rays were first observed by Plücker in 1859, and in Germany they were generally considered to be waves, whereas in England the view was held that they were particles—in fact, that they were charged atoms or molecules.

A new era in particle physics began in 1897 when W. Kaufmann, J. J. Thomson and Wiechert measured the charge-to-mass ratio of the cathode ray. The result of Thomson's work led to the conclusions (1) that the atoms are not indivisible, for negatively electrified particles can be torn from them by the action of electric forces, by the impact of rapidly moving atoms and by ultra-violet light and heat, (2) that these particles are all of the same mass, carry the same charge of negative electricity from whatever kind of atom they may be derived and are a constituent of all atoms, and (3) that the mass of one of these particles is less than one thousandth part of that of an atom of hydrogen.

Thomson originally described these particles as *corpuscles* but, with the adoption of the name given by Johnstone Stoney to the amount of electricity required in electrolysis to liberate one atom of hydrogen, they were soon known as *electrons*. The first direct measurement of the electronic charge was made by Townsend in 1897.

At the same time a study of rays travelling in the opposite direction in the discharge tube proved the existence of positively charged masses, but the unravelling of their nature did not prove so simple. They consist of atoms and molecules of residues of gases present in the discharge tube. The positive charges are found to be equal in magnitude to the electronic charge or small integral multiples of it. The difficulty in determining the charge-to-mass ratio of these particles was overcome by J. J. Thomson who caused the particles to enter a region in which there were parallel electric

and magnetic fields. The particles enter the region at right angles
to the fields and after leaving it fall on a photographic plate placed
at right angles to their original direction. On developing the plate
parabolæ are found upon it. Particles with the same charge-to-mass
ratio lie on the same parabola so that it is possible to sort out different
values of the ratio for the particles in the discharge. A particular
value of charge/mass found in this way is equal to that for the hydrogen ion in solution. Thus this particular particle is identified with a
hydrogen atom carrying a single charge.

By refinements of the apparatus Aston has shown that every
particle registered in this way on a photographic plate has a mass
which can be measured by a whole number when oxygen is taken
as 16. This gives strong proof that matter is built up of fundamental
particles which are called *protons*, and thus restores Prout's hypothesis.

Electromagnetic theory of matter

In this discussion there has been no more than a suggestion of
the intimate relation between electricity and matter. The discoveries described suggest that both electricity and matter are built
up of individual units. The question may clearly arise whether this
property of atomicity is simply a characteristic which happens to be
shared by the two substances or whether this common feature denotes
a closer union between electricity and matter.

This question brings us to the electromagnetic theory of matter.
The electrons in a discharge tube can be bent from a straight path
by the application of electric and magnetic forces. This indicates
that they have an inertial property and so it is natural to associate
mass with them. The question is, are we concerned with electricity
attached to ordinary matter, as we suppose is the case in electrolysis,
or are we to associate with electricity itself the property of inertia?
That is to say, need the particle be any more than a particle of
electricity? This is clearly a point of great interest and importance,
for if it can be deduced that electricity by its nature has the property
of mass then an electromagnetic theory of matter is in sight or is at
least suggested. Towards the end of the nineteenth century speculations on this question were in the forefront of inquiry in physics.
Two names must be mentioned in this connection—Abraham and

Lorentz. There is an important difference in some of the detail of their work but the underlying idea is the same. Their object was to show that the field generated by a charged particle in motion reacted upon it and that the energy of the field of the particle represented the whole of its energy. The field was also considered to possess electromagnetic momentum which was to be identified with the mechanical momentum of the particle.

This was an attempt to relate all the mechanical properties of the electron to electromagnetism, but the attempt did not prove successful. If the momentum and hence the mass of the particle be considered as wholly electromagnetic in origin, the electromagnetic energy turns out to be too small. It is necessary to recognize the existence of energy of non-electromagnetic character. The inadequacy of the purely electromagnetic theory is also apparent in the appearance of a length described as the radius of the electron. It has not proved possible to account for this length in a satisfactory way; it enters with the assumption that the electron is a charged sphere of a certain radius. The value of this fundamental length, which is of the order of 10^{-13} cm., is found by experiment.

There is, however, no fundamental difficulty associated with the introduction of this length in the classical theory of the electron. In fact a finite size is required to avoid the infinite value of the energy of the electron, which would result from assuming it to be a point charge. The need for a non-electromagnetic contribution to the energy of the particle becomes clear from the assumption of a finite size of the particle since the tendency of a body, electrically charged, to disrupt by the mutual repulsion of its elements must be balanced by some force holding it together.

Owing to the non-invariant character of this length in the theory of relativity the possibility of setting up a satisfactory invariant theory of the electron presents difficulties. This is especially acute in the quantum theory of the electron where it has been considered necessary to regard it as a point charge, thus introducing infinities which in general are only avoided by disregarding them. There are further difficulties in the quantum theory traceable to the endowment of the particle with some form of structure, and the tendency is to look for a theory avoiding any reference to the structure of the particle. Work still goes on in the attempt to set up a satisfactory classical theory of the electron in order to pass from it to a satisfactory quantum mechanical theory.

Fundamental Particles

Thus in the early years of the present century it was generally held that there were two fundamental particles, one comparatively massive carrying a positive charge and the other relatively light carrying a negative charge.

Of the particles observed in radio-active phenomena, the α-particle turned out to be the helium nucleus and was regarded as a structure of four protons held together by two electrons, while the β-particle was identified with the electron (see p. 49-50).

In spite of the inadequacy referred to previously, it was still hoped that an electromagnetic theory of matter might ultimately be developed, the fundamental entity in the universe being electricity in the two forms. If this theory could be satisfactorily developed all the forces of nature could be recognized as electromagnetic in origin and the laws of mechanics would ultimately be laws of electromagnetism. The theory of matter lasting until 1932, based largely on the work of Rutherford, was that atoms consisted of nuclei surrounded by electrons moving in orbits and the nuclei themselves were thought to consist of protons and electrons. The scheme of things was thus very simple. All the variety of inorganic and organic matter was the variety of arrangement of electrons and protons. The symmetry of the scheme was marred by the asymmetry in the nature of the particles themselves, the positive being about 1840 times as massive as the negative. This theory of matter met with extraordinary success—it inspired research and in a large degree satisfied the requirements of experiment for a considerable period. The electromagnetic theory of matter, based upon Maxwell's equations, however, made no further progress and the difficulty it raised has not yet been satisfactorily overcome.

The success of the two-particle theory in the explanation of the known experimental facts of atomic physics can hardly have been surpassed at any time. Nothing before it had exceeded it in accuracy both in explaining facts already known and in foretelling much that remained to be discovered. Its outstanding success following the revolutionary ideas of Bohr lay in the field of spectroscopy, where it provided a rational explanation of the numerous empirical regularities known to spectroscopists.

The New Mechanics

But it has to be admitted that the Rutherford and Bohr model of the atom contained the seeds of its own destruction. It introduced quantum principles into atomic mechanics and led to the development of quantum mechanics which, for the present at least, has put aside models, offering in their place another powerful instrument to theoretical physics—the appeal to certain mathematical forms. Nevertheless until about the year 1925, in spite of difficulties in the use of the model to explain the interaction of the atom with a magnetic field, there was still a general belief that progress would continue by this well-established method. Its latest success had been a very remarkable one. The introduction by Goudsmit and Uhlenbeck of the concept of an intrinsic angular momentum, i.e. a rotation or spin of the electron, led to a great and final triumph for the model. The work of L. de Broglie, Dirac, Heisenberg and Schrödinger in the years 1924–5 brought a new method of approach to the solution of the problems of atomic physics. De Broglie and Schrödinger laid the foundation of the new theory by making use of concepts associated with wave motion. Their work revealed a wavelike characteristic of a particle, and the theory developed from it is described as *Wave Mechanics*. Heisenberg and Dirac kept more closely to the Hamiltonian form of particle dynamics, and although the notation employed differs from that of classical mechanics they made an appeal to the mathematical forms of that theory and developed the subject of *Quantum Mechanics*. It was soon discovered that these apparently widely differing lines of approach to the problem could be united, and Schrödinger showed the relation between the two points of view.

It was a consequence of de Broglie's work that a wave must be associated with a particle. A quantitative expression of this association requires a relation between a quantity peculiar to the wave and one peculiar to the particle. The related quantities are wavelength (λ) and momentum (mv) and the relation is $\lambda = h/mv$, where h is Planck's constant (see p. 000). This result has in the past twenty years received ample experimental verification. This was the beginning of the association of a wave-like character with a particle, but twenty years earlier a particle-like property had been associated with a wave. In 1905 Einstein made the suggestion that radiant energy was not only emitted in quanta but also absorbed in these elements which are now known as *photons*. The prediction which follows from Einstein's equation was accurately verified by Millikan ten years later.

The particle-like property of these elements of radiant energy is illustrated in a certain type of X-ray scattering known as the *Compton effect*. It had been observed by a number of physicists that X-rays after falling upon matter gave rise to secondary X-rays of greater wave length. In 1922 A. H. Compton made a spectroscopic investigation of the secondary X-rays from light elements and gave a quantitative explanation of the observations by treating the phenomenon as a collision process between particles. The photon is considered as a particle possessing energy and momentum which strikes a free or loosely bound electron. By making use of the principles of conservation of energy and momentum it can be deduced that the photon will be deflected with less momentum and therefore with an increased wave-length. The formula for the change of wave-length obtained in this way is in agreement with experiment and very interesting in its form. The change in wave-length depends only on the angle of scattering and not upon the nature of the scatterer. It introduces the length h/m_0c, known as the *Compton wave-length*, which is of fundamental importance in physics, m_0 being the mass of the electron and c the velocity of light.

It is interesting in looking back to notice that the associations of wave with particle and particle with wave might have been expected to have occurred together, for according to the principle of relativity a momentum-energy vector is associated with one which is composed of wave-number $(1/\lambda)$ and frequency. Einstein's proposal concerned the association of energy with frequency, the complementary relations express the association of momentum with wave-length— de Broglie's proposal.

The Neutron

The fundamental simplicity of a physical world composed of protons and electrons was disturbed in 1932 by Chadwick's discovery of a third particle, the *neutron*, which carried no charge. The discovery was preceded by Bothe and Becker's experiments which showed that the bombardment of the lighter elements, in particular boron and beryllium, by alpha particles resulted in the generation of a penetrating radiation. The work of F. Joliot and I. Curie showed that this radiation could eject protons from material containing hydrogen, for example paraffin wax, with very great energies. It was suspected that the penetrating radiation was of the nature of gamma rays, but the attempt to interpret the observations on this basis led to inconsistencies and indicated to Chadwick that the radiation consisted of uncharged material particles of about the same mass as that of protons. The fact that these particles are uncharged means that they are little affected by charged particles and they can interact with nuclei more effectively than protons or alpha particles. They are, however, very small and a direct hit is a very rare event. They make no track in an ionization chamber, as charged particles do, and reveal their presence only by occasional collisions with atomic nuclei which then produce ionization. It is from the measurement of the momentum and energy in the nuclei resulting from a collision that the mass of the neutron has been determined.

The Positron

In the year following the discovery of the neutron another particle was discovered by Anderson. This was a positively charged particle of the same mass as the electron, and is the positive counterpart of this well-known particle. It is now generally known as the positive electron or *positron*. This particle had in a certain sense been foretold by Dirac in a study of the solutions of the fundamental equation of quantum mechanics. The best known solution of this equation is that which describes the properties of the negative electron, but there is another associated with it which describes something foreign to the classical and older quantum theory. It was at first thought that the solution referred to the proton, but there was no way of accounting for the asymmetry between the masses of the proton and electron. The discovery of the positron removed this difficulty and revealed the significance of Dirac's discovery.

The experimental discovery of the positron was by means of the cloud chamber (see p. 51). Anderson observed tracks in pairs which seemed to start from the same point. The tracks showed the same general features but in a magnetic field they curved in opposite directions. This showed that one was negatively and the other positively charged. The tracks themselves suggest that the particles forming them are alike except for the sign of the charge and, in particular, that the masses are the same. Later measurements of the mass of the positron by a method similar to that by which the mass of the electron has been determined have confirmed this equality.

The fact that the tracks start from the same point suggests that the electron and positron are generated together. It is in agreement with the theory of Dirac mentioned above that the creation of such pairs of particles should occur. Energy in the form of radiation can produce the pairs of particles, and pairs of particles can unite to produce radiation. The passage of γ-rays through matter produces radiation by this process of pair production. In a strong electric field, such as that near an atomic nucleus, it appears that part of the energy of the γ-ray is converted into a pair of particles and the energy left over appears in the form of the kinetic energy of the particles. In the reverse process the destruction of a positron can occur by union with a free or with a loosely bound electron. Processes of this kind in cascade have been suggested in explanation of the shower process in the cosmic-ray phenomenon. A high energy photon passing near a nucleus undergoes the process of pair production. The charged particles thus produced undergo acceleration in the fields about the nuclei and radiate new photons. These again give rise to new pairs and so on until there is insufficient energy left to carry on the process. This results in an extended group of many particles known as a shower, which may be observed in cloud chambers.

Structure of the atomic nucleus

The development of the wave-mechanical and quantum-mechanical theory revealed a serious difficulty concerning the idea that the atomic nucleus is made up of protons and electrons. According to this theory it is impossible to think of an electron as existing within a region of the dimensions of the nucleus. It is now supposed that atomic nuclei are composed of protons and neutrons and that the stability of the structure is maintained by a force between these two particles, not a force of electromagnetic character such as would be supposed to exist between the charged protons, but a new type of force which brings about a transformation of proton into neutron or *vice versa*. These two particles are regarded as being different states of a single particle known as a *nucleon*. As in the case of the electrons surrounding the nucleus in an atom a transition from one orbit or state to another is associated with the emission or absorption of energy, so the nuclear transformation of proton to neutron, or inversely, is to be expected to give rise to emission and absorption.

This theory of nuclear structure has been very successful in throwing light on the question of nuclear energies and in revealing new facts, while the analogy with atomic transitions has been responsible for some success in developing a theory of the nucleus.

The Neutrino

The α- and β-particles emitted in natural radio-activity have already been mentioned. In 1934 Curie and Joliot discovered the phenomenon of artificial radio-activity as a result of the bombardment of aluminium with α-particles. It has been found that positrons as well as electrons are emitted in these artificial radio-active transformations. In those occurring naturally the positive particle has not been found. The new quantum mechanics is able to explain the phenomenon of α-radio-activity. The nature of β-radio-activity has, however, raised a difficulty, for it has been well established by experiment, in particular by Ellis, that β-rays are emitted with a continuous range of energy values. The difficulty is that on the supposition that the particle is emitted by the nucleus, the latter is left in a different energy state according to the energy of the particle emitted. Now one β-particle is emitted when a nucleus is transformed, so that it appears that the atoms of a substance can undergo the same type of transformation and be left with different energies. This result seems at variance with the principle of conservation of energy and it was suggested that possibly this principle did not hold in β-ray transformations. But the principle is, so far as is known at present, universal in the physical world. It has been found valid on the large scale and upon the small scale and it would be surprising if β-radio-activity formed a single exception. To avoid this difficulty it has been suggested that the process consists in the emission of two particles, the electron (β-particle) and a very light neutral one known as the *neutrino*. Efforts have been made to obtain experimental evidence of its existence, but the study of the theory of β-ray spectra based on the emission of the two particles shows that this particle is necessarily of extremely small mass so that, having no charge and negligible mass, it must be difficult to observe even if it exists.

At present it must be regarded as the carrier of energy which, added to that of the β-ray, leaves the atom in a particular state. The problem of β-ray emission is then that of finding how a given amount of energy is shared between the two particles, based on the principle of conservation. There is also the equally important question of conservation of angular momentum which is violated if only the emission of an electron is contemplated in the process. A particle with a spin angular momentum equal in amount to that of the electron is required to save this principle also.

This particle must, until more definite evidence of its existence becomes available, be regarded as hypothetical.

Fermi's Theory

A great advance in the realm of nuclear physics was made when Fermi undertook the development of a theory of β-ray emission. He based his theory upon the idea that the emission consisted in the change of state from a neutron to a proton combined with the emission of an electron and a neutrino. This seems to imply the existence of a reverse process in which a positron and another particle called the *antineutrino*, are emitted. An appeal is then made to the form of the theory of the atomic emission of photons. Fermi's theory had some success in explaining the character of the β-ray spectrum but there is an implication in the analogy with photon emission in atoms which gives rise to a difficulty.

The β-particle and neutrino are to be compared with the photon. The latter is associated with an electromagnetic field of which the electrons are the sources. Thus the particle and neutrino would be expected to be associated with a nuclear field of which the nucleon is the source. The quantum field theory makes it possible to refer the Coulomb force, i.e. the ordinary force of attraction or repulsion between electric charges, to an exchange of photons between electrons. This exchange is the expression of the existence of the Coulomb field.

Thus Fermi's analogy should lead to an expression of the nuclear field in terms of interchanges of the electron and neutrino between nucleons. Unfortunately this idea leads to a value for the nuclear binding energy which is far too small. Nevertheless this theory is the starting point of the nuclear theory which, although not yet successful, has been very fruitful. The interest here is that the failure of the theory led Yukawa, in his attempt to overcome the difficulty of accounting for the β-ray spectrum and nuclear binding, to foretell the existence of a new particle, now called a *meson*, about a year before its discovery by Anderson.

Yukawa saw that if it were assumed that the transition from the proton to the neutron state and *vice versa* were associated with the emission of a particle heavier than the electron, the difficulty concerning the binding energy could be overcome. He found that the mass of the particle required was about two hundred times that of the electron. The particle was supposed to have a short life and then to decay into an electron and a neutrino. The character of the β-ray spectrum was successfully accounted for in this latter process. This is another example of the prediction of a new fundamental particle, comparable with that of the positron.

Cosmic Radiation

Anderson's discoveries of the positron and meson were made in the course of an investigation by the cloud chamber of the nature of cosmic radiation.

At the beginning of this century, owing to the work of C. T. R. Wilson and others, it was known that a carefully insulated electroscope, even when surrounded by thick metal screens, slowly discharged with time. It was thought that this was due to ionizing radiations from radio-active substances always present in the earth's crust. A Swiss observer, Gockel, in 1910 sent up an electroscope in a balloon to a height of over 12,000 ft. and observed a slight increase of discharge with height. Hess demonstrated most clearly that the rate of discharge increased with altitude, making measurements up to about 15,000 ft. Kolhörster in 1914 showed that at 27,000 ft. the rate of discharge was thirteen times that at sea level.

Self-registering machines were first used by Millikan and Bowen and records were obtained up to a height of about 45,000 ft. Regener's observations have pushed the height to 28 kilometres, i.e. to a region where 98% of the atmosphere is below the region of observation. Thus it was concluded that the discharge was due to a highly penetrating ionizing radiation having its origin beyond the atmosphere. The radiation was therefore called cosmic.

After the First World War the study of cosmic rays was resumed, and Millikan's observations in deep lakes showed that the absorption corresponds to that of a heterogeneous radiation. It was found that a hard component was observable to a depth of about 1,500 ft. of water. The behaviour with regard to absorption showed a similarity to that of γ-rays so that the early idea of the nature of cosmic radiation was that it consisted of very energetic photons.

It was, however, observed in 1927 that the intensity of the radiation was about 10% less near the magnetic equator than at high latitudes. Thus the suggestion is that the radiation is affected by the earth's magnetic field and that it consists, to some extent at least, of electrically charged particles. It was also discovered that more particles arrive from a westerly than from an easterly direction. This gives a reason for concluding that the particles are predominantly positively charged at ground level.

Tracks of the particles constituting cosmic radiation have been observed in cloud chambers and more recently by means of photographic plates. These plates carry an emulsion especially sensitive to the particles, and the ionization track which the particles produce in it can be developed and studied under a high power microscope.

In the case of a particle moving in a cloud chamber the curvature of the track in a magnetic field gives the momentum of the particle and the density of ionization along the track leads to an estimate of its velocity. From these quantities the mass can be deduced. This, however, is possible only for comparatively slowly moving particles because for rapidly moving particles the ionization is almost independent of the velocity. But in the case of the fast particles the same method at least gives an upper limit to the mass.

The method employed to determine the mass by means of photographic plates is to examine the tracks terminating in the film. The range beyond a particular point is measured and also the grain density in the neighbourhood of the point. The range gives the energy of the particle at the point considered and the grain density gives the velocity.

The study of tracks by these and other methods has shown that at ground level positrons, electrons, protons, positive and negative mesons, neutrons and photons all occur in cosmic radiation. These are almost all secondary particles and the nature of the primary radiation incident upon the earth's atmosphere is not completely understood. The primary radiation is of very high energy, of the order of 10^{10} electron-volts, far greater than that of any fundamental particle previously known. It is thought to consist mainly of high energy protons, although nuclei heavier than hydrogen have been observed.

Mesons

After a long period during which it appeared that only two fundamental particles existed, the discovery of new ones was an event in atomic physics comparable with the discovery of new planets in astronomy. There has thus been a concentration of interest on the tracks of ionizing particles with a view to determining the masses of the particles.

It is now certain that particles exist with masses lying between those of the electron and the proton. These particles, which may be positive negative or neutral, are all called *mesons*. Experiments indicate the possibility that a variety of mesons exists. Of these it can be said with certainty that there are two mesons, each occurring positively and negatively charged, with masses about three hundred and two hundred times that of the electron respectively. The former are known as π-mesons, the latter as μ-mesons. The π-mesons interact strongly with the nucleons of the atom and resemble Yukawa's theoretical particle. On the other hand, the μ-mesons can have no such strong tendency to interact. Tracks of the particles seen in emulsions illustrate this difference, for the negative π-meson near the end of its range is easily absorbed by the nuclei with the formation of a star of particles showing that a disruption has occurred. No corresponding effect is observed in the case of μ-mesons except perhaps very rarely.

Much of the work on this subject has been carried out in Bristol under the direction of C. F. Powell, who has developed the technique of the photographic method and has been largely responsible for the interpretation of the observations. The workers there have shown that the π-meson can decay spontaneously into a μ-meson, the life of the π-meson in this process being estimated to be of the order of 10^{-9} sec. Powell has also developed a method of determining the mass of the particles by counting the number of deflections along the track due to interactions with the atomic nuclei of the emulsion.

In Berkeley, California, π-mesons have been produced artificially and the mass measured accurately. It has been found that the masses of the π- and μ-mesons are respectively 285 times and 212 times that of the electron, with a possibility of error of about 3 per cent.

It is interesting to note that the observation of the decay of π-mesons suggests that there is a release of energy which may mean the release of a neutral particle. This, it has been suggested, is the neutrino. Information concerning the other mesons that have been suggested is less reliable. They can be described as heavy and light. The former have been called τ-mesons, the latter λ-mesons. Their masses are said to lie within the range 800–1000 times and 10–30 times that of the electron, respectively. What is now required is more evidence concerning the existence and properties of these particles and a satisfactory theory of nuclear structure.

The past hundred years has thus seen not only a vast wealth of new discovery but a great change in outlook. The first half of this period produced such advances and such success that by the end of the nineteenth century it was generally held that the problems of natural philosophy had been solved in principle. The first half of the present century has changed this belief. Matter itself is no longer permanent; it may be changed to other forms of energy, particles and waves are not to be sharply distinguished, momentum and position are not simultaneously measurable and the older mathematical techniques are not sufficient to express our new theories.

But in spite of the complexity which new discoveries, like the discovery of new particles, and unfamiliar mathematics appear to have introduced, there remains a strong conviction that there is an underlying unity and simplicity in physical phenomena. In this belief physicists of a hundred years ago and of today are at one.

CHAPTER IV

THE STRUCTURE OF THE ATOM

by W. WILSON, D.SC., PH.D., F.R.S.

ATOMS, as the chemists of 1851 thought of them, were still very little different from those of the original theory (*c.* 1803) of John Dalton. They were visualized vaguely as hard spherical structures and believed to be incapable of sub-division (hence the name *atom*). The chemical phenomena of *valency* already indicated differences between elements, now known to be associated with structural features of their atoms; but the simple notions in the minds of the chemists and physicists of a century ago about the mechanism of valency fell far short of the wonderful reality and such known phenomena as had any relation to an inner structure in atoms could not then be interpreted. The absolute masses of atoms could not be even roughly estimated till about 1865, when the kinetic theory enabled the order of magnitude of numbers and masses of molecules to be computed (Clerk Maxwell, Clausius and Loschmidt). Even the atomic weights of the elements (hydrogen= 1) were not certainly known. Those of oxygen and iron, for example, are given as 8 and 28.04 respectively in Miller's *Elements of Chemistry* (1855), while they are in fact about twice as much.

From the time of the publication of Dalton's atomic theory till the opening years of this century there were indeed chemists and physicists who disbelieved in the actual existence of atoms and molecules, and the first three editions of Ostwald's *Grundriss der Allgemeinen Chemie* dispensed with such seemingly wild hypotheses.

Among the various lines of investigation which led, not only to the conviction of the reality of atoms, but to information about their internal structure, may be mentioned:

(i) the study of the spectra of the light emitted, e.g. by salts volatilized in flames or by gases through which electric discharges were passed;

(ii) the suggestions provided by the periodic recurrence of similar chemical and physical characteristics when the elements were arranged in the order of increasing atomic weights (J. A. R. Newlands, 1864 and D. Mendeléeff, 1869);

(iii) the investigation of the phenomena accompanying the passage of electricity through electrolytes, i.e. through aqueous solutions of acids, salts, etc., and through gases at low pressures—more especially of cathode rays and X-rays and the associated discovery of the *electron*; (originally by Crookes and others in the 1870s and then the actual discovery by Townsend and J. J. Thomson in 1897);

(iv) the investigation of the nature of X-rays (C. G. Barkla) and of the X-ray spectra of elements (H. G. Moseley);

(v) the investigation of radio-active phenomena, especially by E. Rutherford, F. Soddy and their successors; and

(vi) the experimental investigation of *isotopes*, i.e. elements having the same chemical properties but different atomic weights (mass numbers).

Many lines of inquiry of course assisted indirectly in solving problems about atomic structure, notably the investigation of the character of black body radiation which gave rise to the quantum theory (Planck, 1900—*vide* Chapter I).

Spectra

Light is regarded here as a wave propagation. The simplest kind of light is associated with a single wave-length or, more precisely, a very narrow range of wave-lengths, e.g. the yellow light from a bunsen flame in which a sodium salt has been volatilized. Such light is called *homogeneous* (or *monochromatic*).

In the early years of last century Thomas Young and Joseph Fraunhofer had invented simple devices for measuring the wave-length of light. Indeed one of Newton's most famous experiments (Newton's rings) enables this to be done; but his interpretation of it differed from that of Young. Fraunhofer (1814) used a grating of parallel equally spaced fine wires, very close together, and this suggested the later forms of *grating* consisting of parallel equally spaced lines ruled on glass or on speculum metal.

Light from most sources usually consists of a superposition of many homogeneous parts, and the grating separates them and enables their wave-lengths to be determined in a simple way. The ordinary prism spectroscope also separates the homogeneous parts in a beam of light; but does not, without some adventitious aid, enable wave-lengths to be measured. Spectra appear to have been first investigated by Isaac Newton (*Opticks*, p. 18, first edition), who used a glass prism to analyse sunlight.

In the usual forms of spectroscope the light under examination is used to illuminate a narrow slit and the rays of the transmitted light are bent into a parallel beam by a suitable lens. The beam then passes through the prism or other device, which separates or analyses it into a number of homogeneous parallel beams. These are received by a telescope (or camera) which makes images of the slit with the light of the homogeneous beams. The *ensemble* of these images, each of which is associated with a definite wave-length, is called a *spectrum*.

Until the '80s of last century the progress made might, for the most part, be described as finding out the characteristics of the spectra of the light emitted by excited atoms. The wave-lengths of the light producing the spectral lines (slit images) in the case of many elements were determined with considerable accuracy by Ångström (*c.* 1868) and later by Rowland (1883) who made a greatly improved type of grating. Attention began to be given in the '60s to light from stars and nebulæ (William Huggins, 1862; Father Secchi, 1867; and Norman Lockyer a little later). In 1868 the last named observed a spectral line in the yellow region of the spectrum of the sun's outer layers. No terrestrial element was then known to which it could be ascribed and Lockyer had in fact discovered a new element in the sun. He very naturally gave it the name *helium* (ἥλιος = the sun). It was found on the earth many years later (1895) in the mineral *cleveite* and in the atmosphere (W. Ramsay and W. Crookes)

In 1885 a Swiss schoolmaster, J. J. Balmer, made a great discovery. He was obsessed with the pythagorean belief in the importance of whole numbers, especially in the description and elucidation of spectra, and gave his attention to the most prominent series of lines in the spectrum of hydrogen (now named after him, *Balmer's series*). By laborious trial and error he eventually found a quite simple formula which represents the wave-lengths of the lines in this series with considerable accuracy, namely

$$\lambda = 3646 \ \frac{n^2}{n^2 - 4} \ ,$$

in which λ means the wave-length of a particular line in Ångström units (or *angstroms*, as they are now officially called)—one angstrom $= 10^{-8}$ centimetre—and n is any whole number greater than 2. The Swedish spectroscopist Rydberg had a preference for working with reciprocal wave-lengths, i.e. $1/\lambda$, or *wave-numbers* as we say. So Rydberg's way of writing the formula was

$$\text{wave-number} = \ \frac{1}{3646} \ (1 - \frac{4}{n^2}),$$

which becomes, when expressed in reciprocal centimetres,

$$\text{wave-number} = 109700 \ (\frac{1}{2^2} - \frac{1}{n^2})$$

to a close approximation.

Other elements beside hydrogen have series spectra. There are various types of series called *principal* series, *sharp* series, *diffuse* series, etc., from their characteristic appearance; the spectrum of an element may contain series of all types. Their wave-numbers were found to be expressible as differences somewhat in the same way as in Balmer's series, and W. Ritz conjectured that there was a set of *terms* characteristic of each particular atom (and of each molecule too, or of any system which can emit light or radiation) and that the wave-number (and therefore also the frequency) associated with a spectral line was equal to the difference between two of the terms belonging to the atom (or emitting system). This conjecture turned out to be correct and is now known as the *combination principle* of Ritz. The constant, approximately given as 109700 cm.$^{-1}$, is called *Rydberg's constant* and this name is also given to

$$109700 \times 3 \times 10^{10}$$

which replaces 109700 when we wish to calculate frequencies instead of wave-numbers, since

$$\text{frequency} = (\text{wave-number}) \times c$$

and c is the velocity of light in free space, namely 3×10^{10} cm./sec. very nearly.

The bearing of these things on atomic structure will be indicated later.

Rutherford's View of Atomic Structure

Ernest Rutherford (1871–1937), afterwards Lord Rutherford, investigated the scattering of α-particles shot at thin metal foil (gold leaf). These are very small, but rather massive, particles thrown out by radio-active materials. They will receive a fuller description later; meanwhile it may be said that an α-particle has about four times the mass of a hydrogen atom and carries two elementary units of positive electricity (*vide* Chapter III). The scattering was investigated by the aid of the scintillations caused by the particles on striking a fluorescent screen. Some of them, it was observed, were deflected sharply through very large angles and the only way of accounting for this was to suppose that each atom of gold had a massive positively charged nucleus, very small in linear dimensions compared with the size of the atom itself—in fact almost like a massive positively charged point.

Computations, making use of measurements of viscosity and other properties of gases and based on the kinetic theory (Maxwell, *vide* Chapter III), indicated long ago that the diameter of a molecule of, say, hydrogen was of the order of 10^{-8} cm. and that its atoms probably were of this order of magnitude. This has been confirmed since by Bohr's theory. The positively charged nucleus of an atom and also the electron are much smaller in size—10^{-12} or 10^{-13} cm.

Rutherford's experiment suggested to him that probably every atom resembles, on a small scale, the solar system. It has a massive positively charged nucleus, a kind of sun, small in size, indeed almost punctual, with one or more negatively charged planetary electrons revolving about it. This picture of an atom is now believed to be the correct one.

After some earlier and very remarkable work by J. W. Nicholson (1912) in which the first successful attempt was made to calculate wave-lengths from an assumed atomic model, Niels Bohr (b. 1885) a Danish physicist, solved the problem of the structure of the simplest atom, the hydrogen atom, and its line spectrum in 1913.

Bohr's Theory of the Hydrogen Atom

Under the influence of the work of Rutherford and Nicholson, Bohr assumed the atom of hydrogen to consist of a massive positively charged nucleus, now called the *proton*, a sort of sun, with a single (negatively charged) planetary *electron* in motion round it in a circular orbit. Each of the two charges he assumed to be the elementary one, sometimes called the atom of electricity.

The earliest use of the name *electron* was to designate this charge (Johnstone Stoney), in 1874 before it came to be used for a particle. It is often represented by the letter e and is known to be very near to 4.803×10^{-10} electrostatic units.

That the quantum theory was needed in some form was already indicated by the combination principle of Ritz. The existence of spectral terms suggested sudden jumps from one atomic state to another, and Bohr introduced it (1913) in an appropriate and illuminating way which fixed, in terms of a whole number, the possible circular orbits in which the electron was forced to move. He further assumed that when an electron suddenly jumped from one permitted orbit to another (quantum jump) the corresponding amount of energy emitted (or absorbed), when this took the form of light (or radiation of that kind), was equal to the product of Planck's h and the frequency of vibration in the emitted or absorbed radiation. These assumptions accounted in a simple way for Balmer's series and other similar series such as Lyman's, Brackett's and Paschen's series which had already been, or were later, observed.

The value of Rydberg's constant, which Bohr computed from his theory, was also in excellent accord with the value derived from spectroscopic observations, and one of his greatest triumphs was to explain in a satisfactory way the observational fact that the constant has slightly different values for different elements. This emerged from taking the slight motion of the nucleus into account. Bohr's earliest expression for the energy—in which the small motion of the nucleus is ignored—was:

Numerical value of energy $= 2\pi^2 me^4/n^2 h^2$.

This gave the Balmer lines the wave-numbers°

$$1/\lambda = \frac{2\pi^2 mc^4}{ch^3} \quad \left(\frac{1}{2^2} - \frac{1}{n^2}\right);$$

$n =$ a whole number greater than 2,

$e =$ charge on nucleus or electron $(4.803 \times 10^{-10}$ e.s.u.),

$m =$ mass of electron $(9 \times 10^{-28}$ gramme),

$c =$ velocity of light in free space $(3 \times 10^{10}$ cm./sec.),

$h =$ Planck's constant $(6.6 \times 10^{-27}$ ergs $\times$ sec.).

By substituting these values the reader may work out for himself the value of Rydberg's constant and verify that it is approximately 109700 cm.$^{-1}$.

The generalization of the quantum conditions made by Sommerfeld and the present writer endowed an electron orbit of the hydrogen atom with *two* quantum whole numbers instead of the one, n, appearing in the above formula. Strictly speaking, there were three whole numbers, and later a fourth was added, but two of them can be ignored so long as the atom is not subject to external forces. Some of the orbits are circular and some elliptical. One of the two numbers is associated with the radial motion of the electron and the other with its angular motion.

Bohr used the letter n for the sum of these integers and k for the angular one, which, of course, can never exceed n and moreover can never be less than 1. An orbit for which the former was equal to 4, for example, and the latter to 3, he called a 4_3 orbit.

The Atomic Number

The way was already prepared for further probing into atomic structure by the remarkable experimental investigation of the nature of X-rays by Charles Glover Barkla (1877-1944). Among other things he studied the scattering of X-rays by light elements (1911), and the results he obtained indicated that the number of electrons in an atom—it is they which effect the scattering of the X-rays—is identical with the number which gives the position of the element in the periodic classification of Newlands and Mendeléeff. Now since the atom in its normal state is neutral there are evidently just as many units of positive electricity on its nucleus as there are planetary electrons, since each electron carries the elementary unit of negative electricity. The atomic number therefore, which gives the order in which the chemical and spectroscopic characteristics of elements repeat themselves, can be identified with the number of units of charge on the nucleus of the atom. This was brilliantly confirmed by H.G. Moseley (who lost his life in Gallipoli in the First World War). He began his work after the invention by Sir William Bragg (1862–1942) and his son of a device for measuring X-ray wave-lengths, the first X-ray spectroscope, and his conclusions were drawn from the measured frequencies of the characteristic X-radiations which Barkla discovered. The ordering number, called the *atomic number*, is generally, but not invariably, that which represents the order of increasing atomic weights.

The atom of helium has long been known to be about four times as massive as that of hydrogen. Bohr's suggestion for its structure was therefore a nucleus consisting of four protons (the mass of a proton is about 1840 times that of an electron) and since its atomic number is two he assumed two of the elementary charges to be neutralized by two electrons *in the nucleus*. Then, of course, he assumed two planetary electrons. It is now known that the helium nucleus must consist of two protons (charge+2) and two *neutrons*. These latter are very like protons—but are uncharged (J. Chadwick, 1932). Quantum mechanics indicates that it is very unlikely that an atomic nucleus can contain electrons.

In the nucleus of any atom there is the same number of elementary positive charges as the number of protons it contains, the rest of the mass of the nucleus being made up of neutrons. The total number of these elementary particles in a nucleus, protons and neutrons together, is now called the *mass number* of the element and the common name *nucleon* is used for either proton or neutron. The distinction between the meanings of the old term *atomic weight* and the very recent term *mass number* will become clear presently.

Electron Orbits of Atoms

At this point it is well to turn to Newlands' *law of octaves*. The elements, beginning with helium, atomic weight 4, up to and inclusive of fluorine, atomic weight 19 (atomic weights are given here to the nearest whole number, except in the case of chlorine) are arranged in a horizontal line, in the order of their atomic weights (Table I). A new line begins with neon and ends with chlorine. Newlands arranged them in this way, with the exception of helium and neon whose existence was not then (1864) known, or suspected, and he noticed the recurrence of similar characteristics. Neon (just below helium) is an inert gas. Sodium is extraordinarily like lithium and so on. Lastly there is a striking resemblance between chlorine and fluorine.

Bohr conjectured that the helium atom had two electrons in 1_1 orbits. Lithium, atomic number 3, must have three occupied orbits. Two of them he supposed to be 1_1 orbits constituting the same

Table I

2 Helium 4	3 Lithium 7	4 Beryllium 9	5 Boron 11	6 Carbon 12	7 Nitrogen 14	8 Oxygen 16	9 Fluorine 19
10 Neon 20	11 Sodium 23	12 Magnesium 24	13 Aluminium 27	14 Silicon 28	15 Phosphorus 31	16 Sulphur 32 ..	17 Chlorine 35.46

The number above the name of the element is the atomic number; the lower one is its atomic weight.

stable structure which gives helium its chemical inertness. He imagined the third lithium orbit to be a 2_1 orbit. It should be observed that the net positive charge within this 2_1 orbit is just one unit, since the nucleus has a charge $+3$ and there are two negative electrons. Lithium has thus some resemblance to the hydrogen atom. X-ray spectra and other things strongly suggested that all the elements of atomic number greater than that of helium (i.e. greater than 2) possess two electrons in the 1_1 orbits characteristic of the helium atom. They form the K shell of electrons, so-called because Barkla's characteristic K radiation is effected by the expulsion of one (or both) of them followed by an outer electron dropping into its place. The elements from lithium to neon, inclusive, have, in addition to the two 1_1 K orbits, others for which the quantum number n is equal to 2. These are called L orbits because they are associated with Barkla's

L radiation in much the same way as the 1_1 orbits with his K radiation.

Lithium has a 2_1 orbit outside the stable configuration of two 1_1 (helium) orbits. Beryllium has two 2_1 orbits and no atom has more than two n_1 orbits for a given value of n. For a given value of n no atom can have more than six n_2 orbits. Neon has the two 1_1 K orbits of helium, two 2_1 orbits and the complete set of six 2_2 orbits. All these together constitute, like the two helium orbits, an exceedingly stable structure, and neon, like helium, is an inert gas. Sodium, which follows, has the two 1_1 and two 2_1 orbits together with the six 2_2 orbits of neon and a 3_1 orbit outside. Its nucleus with 11 units of positive electricity (atomic number 11), together with the inner system of 10 electrons in orbits for which $n=1$ and 2, constitutes a system with a net charge of $+1$, and this, together with the outer electron, resembles lithium and hydrogen.

This perhaps is the place to point out and emphasize that the chemical behaviour of an element is determined by the system of electrons, or more precisely the outermost shell[1] of electrons, in its atom, and the nature of valency may be illustrated by its simplest type, as it appears in compounds of univalent elements such as HF, LiF, NaF, HCl or NaCl. In the case of sodium fluoride, NaF, for example, the outermost sodium electron fills up, as it were, the place of the missing 2_2 fluorine one—fluorine has only five 2_2 orbits— and thus completes the exceedingly stable system of six 2_2 orbits. In the case of water the electrons of two hydrogen atoms complete the stable system of six 2_2 orbits by occupying the two empty places in the set of 2_2 orbits of oxygen which has only four of the complete set of six 2_2 orbits.

This distribution of different types of electron orbits in atoms was first suggested by E. C. Stoner, now Professor of Theoretical Physics in Leeds, to account for the features of characteristic X-radiation, and independently, on chemical grounds, by the American, Main Smith. The accompanying Table II describes the electron orbits of the atoms up to the inert gas krypton, atomic number 36. The letters K, L, M, N refer to Barkla's characteristic radiations which are associated with the quantum numbers $n=1$, 2, 3, and 4 respectively.

[1]All electrons for which n has the same value are said to constitute a *shell*.

Radio-activity

Our knowledge of the nuclei of atoms is largely derived from the investigation of radio-activity, discovered by Henri Becquerel about 1896. Many elements of high atomic weight,[1] e.g. uranium, radium (first discovered and isolated by Madame Curie) and thorium, make the air round about them a conductor of electricity. They blacken photographic paper in their neighbourhood from which light is excluded and maintain themselves at a temperature slightly in excess of that of the surroundings. This is associated with the emission of radiation which, in naturally occurring radio-activity, was found to be of three types: α-rays, found to consist of positively charged particles which were later identified with the nuclei of helium atoms; β-rays, consisting of high velocity electrons; and lastly, radiation known as γ-rays, which is identical with X-radiation. The information furnished by natural radio-activity has been greatly supplemented firstly by using the naturally occurring fast moving α-particles to bombard normally inactive substances, e.g. nitrogen, and secondly by endowing charged particles, such as protons, with high velocities by artificial means, for example by the *cyclotron* (E. O. Lawrence).

This apparatus enables a succession of impulses to be given to a charged particle which at first may be moving quite slowly. It thus acquires a high velocity and, when used to bombard atoms, is able to bring about their transmutation. Thus high velocity protons striking an ordinary lithium nucleus attach themselves to it and cause it to split into two helium nuclei. The lithium nucleus has a mass number 7 and there are 3 protons in it. The bombarding proton adds 1 to both numbers—making a total mass number 8 and 4 protons. Thus we see how the two helium nuclei are formed, each with mass number 4 and 2 protons.

The evidence that this kind of thing really happens is provided by C. T. R. Wilson's cloud chamber, the water vapour within which (at the moment when an observation is made) is super-saturated as the result of a sudden (adiabatic) expansion. The flying

---

[1]In the earlier days of the atomic theory, *atomic weight* meant the weight of an atom as determined by chemical means, that of the hydrogen atom being taken as the unit. In more recent times the unit was fixed by taking that of oxygen to be 16.

E

α-particles and protons disrupt gas molecules in the chamber and drops of water condense round the fragments. The tracks of the α-particles (or other fast moving charged particles) are thus indicated by lines of fog (lines of water drops) and the track of the proton ending at a lithium nucleus and those of the emerging α-particles can be seen. Another interesting type of apparatus is the counter invented by H. Geiger which enables charged particles (α-particles, protons, etc.) to be counted, without exhibiting visible tracks.

Isotopes

A great deal has been learned about atomic nuclei and *isotopes* from F. W. Aston's experiments with his mass spectrograph in the Cavendish Laboratory. In this apparatus an electric and a magnetic field are simultaneously applied in an ingenious way to make all similarly charged nuclei, which have the same mass number, arrive at the same point or same short line like a spectrum line, on a photographic plate, whatever their speeds may happen to be. It enables different mass numbers to be distinguished and estimated with remarkable accuracy. Thus Aston showed that chlorine was a mixture of two things having like chemical properties, but different mass numbers, 35 and 37, three of the former to one of the latter, which explains the atomic weight 35.46. It is these different varieties of the same chemical element which are called *isotopes* (ἴσος, same and τόπος, place), a name due to Soddy.

There are three known isotopes of hydrogen: ordinary hydrogen with 1 proton as its nucleus; so-called heavy hydrogen (deuterium) with 1 proton and 1 neutron making up its nucleus and tritium with a proton and 2 neutrons in its nucleus. Uranium has three known isotopes. Its atomic number is 92. There are in fact 92 protons in the nucleus of the uranium atom; but the common uranium has a mass number 238 and the other two isotopes have mass numbers 235 and 234. Uranium 238 emits α-particles naturally and in consequence its atom changes into one of mass number 234 and atomic number 90—an isotope of thorium. The end result of the complicated sequences of naturally occurring radio-active transmutations appears to be one or other of the several isotopes of lead.

Transmutation of Elements

The earliest artificial transmutation was effected by Rutherford, who bombarded nitrogen with α-particles. The result was the formation of an isotope of oxygen and the ejection of protons. It was exhibited later in the cloud chamber by P. Blackett and is now indicated in the following way:

$$_2He^4 + _7N^{14} = _8O^{17} + _1H^1.$$

Mass numbers are placed above and nuclear charges below.

Neutrons are liberated when such elements as lithium or beryllium are bombarded with α-particles, thus

$$_4Be^9 + _2He^4 = _6C^{12} + _0n^1.$$

In this case the emission of a neutron (charge 0 and mass number 1) is accompanied by the formation of a carbon atom. The products of many of these artificial transmutations are radio-active (induced radio-activity) and some of them emit *positrons*, particles exactly like electrons except that their charge is positive. Positrons, neutrons and *mesons* (mentioned again below) have been observed also in the *cosmic radiation* which reaches us from interstellar space. The apparent violation of the principle of conservation of energy (*vide* Chapter I) during the emission of β-particles has made it probable that this emission is accompanied by that of energetic uncharged particles called *neutrinos* (Fermi). Our belief in their existence has received some support in recent times from experiments of J. F. Allen (1942).

The question arises: how are the 92 mutually repelling protons in the uranium atom held together, or for that matter, the protons in any atomic nucleus? It is believed that in an atomic nucleus there is a field of force (*mesonic field*), not of the electrical kind, superposed on the repelling electrical one and counteracting it. This field is associated with the transference of a strange particle, the *meson* (*vide* Chapter III) from a neutron to a proton and from a proton to a neutron. In this ball game neutrons and protons exchange their roles, and the resulting attractive force between them is therefore called an *exchange* force. It is very powerful within the limits of a nucleus (10^{-12} cm.) but falls off much more rapidly than the inverse square electrical one. If a massive nucleus such as that of the uranium atom is disturbed and set in oscillation, parts of it are liable to project beyond the bounds of the restraining mesonic forces and to come under the unrestrained influence of the disruptive electrical forces, and the nucleus may split with great violence into two comparable portions (nuclei of barium and krypton). This is part of the secret (if there is now any secret) of the atom bomb. Neutrons, being uncharged, can easily approach charged nuclei, and the naturally provided neutrons of cosmic radiation enter the uranium (235) nuclei and set up the oscillations which bring about their fission. The splitting nuclei in their turn emit neutrons and if these are not moving too fast and the lump of uranium 235 is big enough the *chain reaction*, as it is called, causes the whole lump to explode within a very small fraction of a second.

It is explained in Chapter I that the formation of helium nuclei from protons is associated with the emission of energy, the *binding energy* of the helium nucleus. Protons, deuterons, etc., can only collide with one another when their velocities are very great, because of their repelling positive charges; but this high velocity might conceivably be given to the deuterons of heavy water if they were subjected to the high temperature within an exploding uranium or plutonium bomb. Protons (in the hydrogen of ordinary water) are probably unsuitable because of their propensity for capturing the vital neutrons which bring about the fission of the uranium (or plutonium) nucleus. Thus a new bomb is suggested, the appallingly destructive potentialities of which can be realized by referring (Chapter I) to the energy liberated by the formation of even one gramme of helium.

Cosmical Speculations

Even atoms have their cosmical significance. Eddington convinced himself (on theoretical and very speculative grounds) that he knew the precise number of particles (nucleons) in the universe—about 10^{80}—and it is remarkable that the estimate made on the basis of Einstein's (very different) cosmological theory is quite near to Eddington's number. Lastly there is a strong suggestion of a complementary universe in which all the nuclei are negatively charged and electron orbits replaced by orbits occupied by positrons—a strange image of the universe which is actually known to us.

Table II. Electron Orbits of Atoms.

Element	K 1_1	L 2_1	2_2	M 3_1	3_2	3_3	N 4_1	4_2
H 1	1							
He 2	2							
Li 3	2	1						
Be 4	2	2						
B 5	2	2	1					
C 6	2	2	2					
N 7	2	2	3					
O 8	2	2	4					
F 9	2	2	5					
Ne 10	2	2	6					
Na 11	2	2	6	1				
Mg 12	2	2	6	2				
Al 13	2	2	6	2	1			
Si 14	2	2	6	2	2			
P 15	2	2	6	2	3			
S 16	2	2	6	2	4			
Cl 17	2	2	6	2	5			
A 18	2	2	6	2	6			
K 19	2	2	6	2	6		1	
Ca 20	2	2	6	2	6		2	
Sc 21	2	2	6	2	6	1	2	
Ti 22	2	2	6	2	6	2	2	
V 23	2	2	6	2	6	3	2	
Cr 24	2	2	6	2	6	5	1	
Mn 25	2	2	6	2	6	5	2	
Fe 26	2	2	6	2	6	6	2	
Co 27	2	2	6	2	6	7	2	
Ni 28	2	2	6	2	6	8	2	
Cu 29	2	2	6	2	6	10	1	
Zn 30	2	2	6	2	6	10	2	
Ga 31	2	2	6	2	6	10	2	1
Ge 32	2	2	6	2	6	10	2	2
As 33	2	2	6	2	6	10	2	3
Se 34	2	2	6	2	6	10	2	4
Br 35	2	2	6	2	6	10	2	5
Kr 36	2	2	6	2	6	10	2	6

CHAPTER V

THE STRUCTURE OF MOLECULES

by J. R. PARTINGTON, M.B.E., D.SC.

IN 1850 chemistry was a large and well-developed science. The *Handbook of Chemistry* of Leopold Gmelin, which appeared in an English translation in 1848–72, occupied nineteen volumes, and dealt with theoretical and experimental chemistry as known at the time. A large proportion of the non-metallic and metallic elements had been discovered and many of their important compounds were known. The chemistry of carbon, which was then and still is called organic chemistry, because carbon compounds occur in vegetable and animal organisms, occupied twelve of Gmelin's volumes and hence was then, as now, the bulkiest part of the body of chemical knowledge. Chemical technology was well developed, the scientific side of the processes being well understood, and the applications of chemistry in medicine and the arts had assumed substantial proportions. At the beginning of the period which we are more particularly to survey, chemistry had long passed its primitive stage. In what important respects, then, does the chemistry of 1950 differ from that of 1850? One of the most important of these, in the writer's opinion, is the greatly increased knowledge of the exact way in which the smallest particles of matter of interest to chemists, the so-called chemical molecules, are built up of the still smaller parts, the atoms of the chemical elements of which they are composed. The way in which this more exact knowledge has been attained is sketched in the present section.

The Composition of Molecules

The chemical atomic theory, which we owe to the genius of the English chemist John Dalton, had been proposed a year or two after 1800. It pictures the chemical elements as composed of exceedingly small particles or atoms, which remain unchanged when they enter into or leave the state of chemical combination, the groups of definite numbers of atoms of particular elements forming the molecules of compounds. Water is composed of the two elements, hydrogen and oxygen, and eight parts by weight of oxygen combine with one part of hydrogen to form nine parts of water. Dalton set the atomic weight of hydrogen, the element with the lightest atom, equal to one. If water contains one atom each of hydrogen and oxygen in its molecule, as Dalton assumed, the formula of water will be HO, and the atomic weight of oxygen will be eight. It was known, however, that two volumes of hydrogen gas combine with one volume of oxygen gas to form two volumes of steam, and this suggested that the molecule of water might contain two atoms of hydrogen and one atom of oxygen, its formula being H_2O. This was assumed by Davy. These two formulae of water were still in use in 1850, and there seemed no certain way of deciding which was correct.

It is noteworthy that the fundamental method of fixing chemical formulae had been laid down by the Italian physicist, Amedeo Avogadro, in 1811, but by one of those curious examples of the neglect of scientific ideas, his publication was not, for a number of reasons which need not be told, given the attention it deserved. It was not until 1858 that the Italian chemist Cannizzaro revived Avogadro's method and made it possible to choose without ambiguity the correct formulae of compounds. Avogadro had assumed that equal volumes of gases and vapours (at the same temperature and pressure) contain equal numbers of particles, but the example of the formation of steam just given shows that these cannot be atoms. The volume of steam is double the volume of the oxygen, and hence the oxygen particle must have been divided into two parts. The oxygen and hydrogen molecules are, in fact, O_2 and H_2, and the water molecule is H_2O. The same result had been reached in 1850 in another way by the English chemist Williamson, who also showed that other compounds are built up on what he called the *water type*. Alcohol and ether have the formulae $(C_2H_5)_2OH$ and $(C_2H_5)_2O$, the ethyl radical C_2H_5, composed of two atoms of carbon and five of hydrogen, taking the place of an atom of hydrogen in water. The recognition of such radicals was very important in the development of organic chemistry, and long before Williamson, Berzelius in

Sweden and Liebig in Germany had shown by chemical evidence that many radicals are present in the molecules of compounds of carbon. About 1850, therefore, the formulae of molecules were fairly well known, and some features of the structure, in terms of the existence of radicals in them, had often been made out.

The Theory of Valency

The atoms in molecules were assumed from Dalton's time to be linked together by what was called *chemical affinity*, which was pictured as some kind of force acting only over a very small distance, and thus different from gravitation, which acts over great distances. Davy and Berzelius had given reasons for thinking that chemical affinity was electrical in character, since many compounds in solution are split up by an electric current. A most important step was made in 1852 by Edward Frankland. He assigned to every atom a number of units of combining capacity, called *valency*. Since one atom of oxygen holds two atoms of hydrogen in the molecule H_2O, the valency of oxygen is two, that of hydrogen being the unit. In hydrochloric acid, HCl, the valency of chlorine is one, in ammonia, NH_3, the valency of nitrogen is three. In 1858, Kekulé, one of the most famous German chemists of the nineteenth century, pointed out that, since the simplest compound of carbon and hydrogen is marsh gas, CH_4, the valency of carbon is four. This valency is shown by carbon in all but a very few of the hundreds of thousands of its compounds. Since the great majority of organic compounds contain carbon, hydrogen, oxygen, and sometimes nitrogen, their formulae must be such as to give these four elements the valencies 4, 1, 2, and 3, respectively.

Frankland also supposed that two atoms in a molecule are linked by a definite directed bond, called a *valency bond*, and each atom has as many bonds as it has units of valency. The four atoms named were thus represented as $=C=$, $H—$, $—O—$, and $=N—$. When they combine, the valency bonds link together, the simple molecules being shown as:

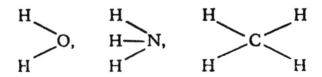

These are called *graphic formulae*. For convenience they are abbreviated by using a dot for a valency bond, and since radicals can take the place of atoms, their valencies can also be found. The formula of alcohol, for example, can be written $C_2H_5 . O . H$, or (what is the same) as $CH_3 . CH_2 . O . H$. Two atoms of carbon may be joined by one, two or three bonds, called single, double, and triple bonds, as in ethane $H_3C—CH_3$, ethylene $H_2C=CH_2$, and acetylene $HC\equiv CH$. Compounds with double and triple bonds are called *unsaturated*, since they easily add other atoms to form single bonds out of the double and triple bonds, such as

$$
\begin{array}{ccc}
Cl & & Cl \\
\diagdown & & \diagup \\
H—C—C—H, \\
\diagup & & \diagdown \\
H & & H
\end{array}
$$

in which the carbon atom has its usual valency of four, each carbon linking two hydrogens and one chlorine, and using its remaining bond to link another atom of carbon. The capacity of carbon atoms to link together into chains was also recognized by Kekulé, and it explains why so many carbon compounds are known.

The Structure of Benzene

The hydrocarbon benzene, C_6H_6, was discovered by Faraday in 1825. By 1850 it was known that many carbon compounds are derivatives of benzene. Phenol, or carbolic acid, for example, has the formula C_6H_6O, and by taking away the oxygen, benzene is formed from it. These substances often have peculiar and sometimes pleasant odours, and are called *aromatic* compounds, another large group of carbon compounds being called *aliphatic* compounds. Whereas the structural formulae of aliphatic compounds can all be represented by the kind of valency bond arrangement shown above, with chains of carbon atoms, this did not prove possible for the benzene derivatives. If we write the formula of benzene with alternate single and double carbon linkages, two valency bonds at the end of the chain are left over, $—C=C—C=C—C=C—$. Kekulé

$$\dot{H}\quad \dot{H}\quad \dot{H}\quad \dot{H}\quad \dot{H}\quad \dot{H}$$

in 1865 solved this problem by assuming that the chain forms a closed ring, the two free bonds at the ends linking together, and so the famous benzene ring came into being:

$$
\begin{array}{c}
H \\
\cdot \\
C \\
H.C \diagup \quad \diagdown C.H \\
| \qquad \| \\
H.C \diagdown \quad \diagup C.H \\
C \\
\cdot \\
H
\end{array}
$$

Since there are so many benzene derivatives, chemists represent the benzene ring by a simple hexagon, the atoms or radicals which substitute atoms of hydrogen in the ring being shown. Phenol, for example, has a hydrogen atom substituted by a hydroxyl radical, —O—H, and its formula is written as 'I' below; aniline has H replaced by the amino-group —NH$_2$, and its formula is shown in 'II':

I. Phenol   II. Aniline

Synthetic Organic Chemistry

Although about 1800 it was thought that the substances in plants and animals were formed by the agency of a special vital force and could not be made artificially in the laboratory, this idea had been quite given up in 1850, since many organic compounds had been made by starting with the elements, or, as is said, by *synthesis*. Acetylene, C$_2$H$_2$, is formed from carbon and hydrogen at a high temperature, and from acetylene alcohol can be made. From alcohol many other compounds, such as acetic acid, can be obtained, and all of them go back to the elements by way of the intermediate steps. When the structural formulae of compounds had been found by their chemical reactions, it was possible to prepare the substances by synthesis. Great numbers of new substances which do not occur in nature can also be synthesized, such as aniline and other dyes, and drugs like aspirin. The second half of the nineteenth century was a period of great activity in synthetic organic chemistry, and chemists of all countries took part in this work. The first aniline dye was discovered by the English chemist, William Perkin, and his son, William Henry Perkin, was an outstanding figure in organic chemistry. In place of the twelve small volumes of Gmelin's book, very large treatises in many volumes are now devoted to organic chemistry, and the number of known organic compounds must now be at least 300,000. The greater proportion of these do not occur in nature at all, but are synthetic products. In the face of this imposing monument to the industry of the synthetic organic chemists we must retreat, and follow another path.

Kekulé pictured the benzene ring as a flat hexagon, for which there was much chemical evidence based on the number of *isomers*, or compounds having the same composition, derivable from it. There was only one nitrobenzene, $C_6H_5 . NO_2$, but three dinitro-benzenes, $C_6H_4(NO_2)_2$, and so on. This picture of the benzene molecule, which guided chemists from 1865 through the maze of aromatic compounds, is now known (as will be mentioned later) to be correct. The isomers in this case depend on the different positions in the benzene ring. It might be supposed that there should be two dinitrobenzenes according as two adjacent NO_2 groups are at the ends of a single or a double carbon bond, but there is only one with adjacent groups. Kekulé, therefore, supposed that the single and double bonds oscillate rapidly in the ring, so that the actual bond between two carbon atoms in the benzene ring is intermediate in character between a single and a double bond, and all the six bonds between the carbon atoms are identical. This picture is also in agreement with modern theory, being related to the so-called resonance effect between bonds of different types.

Stereochemistry

The structural formulae of compounds are drawn on a flat piece of paper, and although the benzene ring is really flat, like its formula, this may not be true of other molecules. The atoms in molecules can be arranged in three dimensions. Thus another branch of chemistry comes into view which is called *stereochemistry*. This had its origin in the work of Pasteur in 1848 on the tartaric acids. There are three isomeric tartaric acids. Two of these, in solution, rotate the plane of polarization of light, and one just as much to the left as the other rotates it to the right. The third is optically inactive. An explanation of Pasteur's work was given independently in 1874 by the Dutch chemist van't Hoff and the French chemist LeBel. The atoms in the two optically active tartaric acid molecules are arranged in space so that one molecule is the mirror-image of the other, like a right-hand glove and a left-hand glove, which cannot be super-imposed on one another facing the same way. This follows if the four valencies of the carbon atom are arranged in space in such a way that if the atom is supposed to be at the centre of an imaginary tetrahedron, the four valencies are directed towards the four corners of the tetrahedron. This so-called tetrahedral carbon atom (it is the valencies which are distributed tetrahedrally, and the name does not imply anything about the shape of the atom) has proved sufficient to explain the optical activity of an immense number of carbon compounds. The carbon atom has the same arrangement of valencies in its other compounds, but in optical isomers there are four different groups or atoms attached to the four bonds. Such an arrangement always exists in two non-superimposable or, as they are called, *enantiomorphous*, forms. Many other elements besides carbon can give rise to optical activity, for similar reasons, and in many cases the absence of symmetry in the molecule, rather than in the bonds of an atom, can give rise to optical activity.

The existence of optical activity gives another important means of elucidating the structure of molecules, and in all cases its results have been fully confirmed by newer methods. Chemists, with their benzene ring and tetrahedral carbon atom, were some decades ahead of the physical methods which must now enter the picture.

Physical Chemistry

Early in the nineteenth century, heat, light, electricity and magnetism were regarded as imponderable (weightless) substances, and they appeared as such in the tables of the elements. Their study formed part of chemistry, such physics as then existed being mostly concerned with mechanics or the study of forces. It was not until about 1850 that it was realized that heat and light are not forms of matter but of energy, and chemists handed over their study to physicists. The first volume of Gmelin's book still dealt almost entirely with what is now the subject of physics.

From 1800 some important branches of chemistry were concerned with the relations of heat and electricity with chemistry. Thermochemistry, with its measurements of the heat evolved in chemical changes, hoped to relate these with the affinities of the combining elements. Electrochemistry, which had given Davy the means of isolating several new elements by the decomposing action of the electric current, had been developed by his pupil Faraday. It had in its time formed almost the sole basis of chemical theory in the hands of Berzelius, whose electrochemical or *dualistic* theory, proposed in 1811, was still in vogue in 1850. It assumed that every compound had an electropositive and an electronegative part, sulphuric acid, for example, being SO_3+H_2O. This theory had declined in organic chemistry, where it had been replaced by one which emphasized the molecule as a whole entity, parts of which could be exchanged by substitution, say of hydrogen by chlorine.

In 1886, Svante Arrhenius in Sweden put forward the theory of electrolytic dissociation, according to which a salt is split up in solution into positive and negative charged ions, which move through the solution with the electric current in electrolysis. Common salt in solution does not exist as NaCl molecules but as the charged ions Na^+ and Cl^-. This theory met with great opposition, but by about 1900 facts had compelled its practically universal adoption. Other branches of physical chemistry, as that part of chemistry concerned with thermochemistry, electrochemistry, etc., was called, were developed, and the whole subject, which is concerned with the application of physical methods and ideas to chemistry, has made great progress since that time. It must be admitted that in Great

Britain, where the main interest was, and still largely is, centred in synthetic organic chemistry, the development of physical chemistry was less significant than in Germany and America, but this defect has been made good in more recent times.

Modern Physics and Chemistry

Although physics and chemistry have always had many interests in common, it is only since about 1900 that discoveries in physics have had a profound effect on chemistry. The discovery of X-rays, of the electron, and of the diffraction of X-rays by crystals, were experimental achievements in physics which had far-reaching applications in chemistry. Radio-activity led to a knowledge of the structure of the atom, and the quantum theory underlies the modern theory of valency. The story of the development of this new physical knowledge is told elsewhere in this volume, and a general acquaintance with its main features is assumed. The following account deals only with those aspects of modern physics which have been used in elucidating the structures of molecules.

The atoms were regarded by chemists as minute spheres, linked together by valency bonds into definite structures to form molecules. Each atom keeps more or less to a fixed place in the molecule, so that a picture of the molecule can be attained if the distance between each pair of atoms and the angles between the valency bonds can be found. It is only rarely that a model made by linking small spheres by rods or wires will be insufficient, but it must be remembered that the atoms have definite sizes, all of the order of 10^{-8} cm., and they may be practically in contact in the molecule, the valency lengths being the distances between the centres of the atomic spheres. Some of the most important methods used in determining molecular structure will now be considered.

X-Rays and Crystal Structure

Since the earlier part of the nineteenth century, crystals have been regarded as ordered structures of atoms called *space lattices*. A crystal of rock-salt was supposed to be formed by piling together atoms of sodium and chlorine into a cubic structure, and it was later recognized that the particles are really the charged sodium and chlorine ions. The whole crystal does not contain any NaCl molecules but only Na^+ and Cl^- ions. The atoms are too small to scatter light waves, which pass them by like a wave on water passing a floating cork. The X-rays, however, have short wave-lengths of about the same size as atoms, and hence they are scattered by atoms. In 1913 diffraction patterns produced by the action of a crystal on X-rays were obtained, and the further study of the subject by Sir William Bragg and his son, now Sir Lawrence Bragg, led to the elucidation of the structures of many molecules in crystals.

The diamond crystal was shown to consist of single atoms of carbon linked by tetrahedral bonds, as had been assumed in aliphatic carbon compounds by chemists. In graphite, another crystalline form of carbon, the atoms are linked in flat hexagonal rings like the benzene ring, and in aromatic compounds such benzene rings are found. The distances between the atoms, i.e. the valency-bond lengths, can also be measured and are all of the order of 10^{-8} cm., the same as the atomic radii. In carbonates, such as Iceland spar or calcium carbonate, $CaCO_3$, the carbonate ion CO_3^{--} is a flat equilateral triangle with a carbon atom in the centre and three oxygen atoms at the corners. In sulphates, the sulphate ion SO_4^{--} is a tetrahedron with a sulphur atom at the centre and four oxygen atoms at the corners. Even quite complicated molecules of organic compounds have had their structures made out by the X-ray method. A more recent modification of the method makes use of the diffraction of beams of electrons instead of X-rays, these behaving as if they had a definite wave-length, and the electron diffraction method, which was developed by Sir G. P. Thomson, son of Sir J. J. Thomson, has played a prominent part in the discovery of the structures of molecules.

Electric Dipole Moments

Some molecules, such as the hydrochloric acid molecule, HCl, are electrically polar, i.e. have equal positive and negative charges separated and forming what is called a *dipole*, which is analogous to a magnet. Such molecules tend to set themselves in the direction of an applied electric field, and by suitable measurements the moment (charge×distance between charges) of the dipole can be found. A dipole moment is a vector quantity; it has direction as well as magnitude, and in combining the moments of the various valency

bonds to form the resultant moment it is necessary to add them like·
forces, which are combined by the parallelogram of forces method.
By suitably choosing the valency angles the correct value of the
resultant moment is found, and hence it is possible to form estimates
of the valency angles.

As an example of the use of dipole moments in deciding the
structure of molecules the case of the three disubstituted benzenes
may be considered. Suppose that two atoms of hydrogen in benzene
are replaced by two atoms of chlorine. Three different compounds
are formed. If we suppose the six carbon atoms in the benzene ring
to be numbered from 1 to 6 in going round the ring, the two
chlorine atoms may be in places 1 and 2, forming orthodichloro-
benzene, in positions 1 and 3, forming metadichlorobenzene, or in
positions 1 and 4, forming paradichlorobenzene. All the other
cases would be the same as these, e.g. 1 and 6 is the same as 1 and 2.
The carbon to chlorine link in the ring has an electric moment. In
the para-compound the two moments are equal and opposite, and
this compound has practically zero resultant moment. The ortho-
compound will have the largest moment, and the meta-compound
a moment of intermediate size. It is,therefore, quite easy to decide
which particular dichlorobenzene is in question by measuring the
dipole moment.

Absorption Spectra

The discovery of spectrum analysis by Bunsen and Kirchhoff in
1859 comes in the period under review. Sir Henry Roscoe, a pupil
of Bunsen, was an English chemist who did notable work in the
chemical applications of spectroscopy, and physicists in Great Britain
have also played a major part in this branch of science. The discovery
of new elements by the spectroscope will be described in the next
chapter. Here the application of absorption spectra to the determination of molecular structure will be considered. This is very intimately connected with the quantum theory.

In 1901 the German physicist Planck proposed the theory that
in the emission and absorption of radiation by matter, the process
does not occur continuously, but the radiation behaves as if its
energy were parcelled out in *quanta*, the size of the quantum being
proportional to the frequency of the radiation; the factor of pro-

portionality is a universal constant called *Planck's constant* and denoted by h. If v is the frequency of the radiation, the corresponding energy quantum is $\varepsilon = hv$. (See Chapter I.)

The simplest molecule is composed of two atoms at a fixed distance apart, the whole system being in rotation. If this molecule has an electric moment, it will interact with infra-red radiation by increase of the rotation of the molecule, and part of the radiation will be absorbed. The missing frequency which has been absorbed is shown by the infra-red spectrum. The molecule when rotating has a definite energy, and by the absorption of a definite quantum of energy its rotational energy is raised to a higher level. If E_r' and E_r'' are the two rotational energies, the difference $E_r'' - E_r'$ is equal to the energy quantum absorbed, hv_r. Since v_r is known from the absorption spectrum, the energy difference can be calculated. This depends directly on the moment of inertia of the molecule, i.e. on the sum of the products of the atomic masses and the squares of the distances of the atomic centres from the centre of gravity of the molecule. The latter distances can be calculated and their sum is the molecular diameter.

Absorption of radiation in the shorter wave-lengths is caused by increase of the vibration of the atoms along the line of their centres, and since rotation is also occurring, the resulting absorption spectrum contains bands, each consisting of a main level due to the vibrational energy transition, $E_v'' - E_v' = hv_v$, with a fine structure of lines much closer together due to the rotational transitions. Finally, there may be larger energy changes due to the absorption of large quanta so as to change the energy of the molecule due to electronic transitions, and the corresponding absorption is in the visible or ultra-violet spectrum. Each electronic level has vibrational and rotational fine structure. *Raman spectra* are due to the scattering of visible light by molecules and change of frequency owing to the interaction of the incident light quantum with the vibrational or rotational energy of the molecule. Such spectra can give much information about the distances of valency bonds in molecules. From them the *bond strengths*, i.e. the resistance of the bonds to stretching, can also be calculated.

If the vibrational energy of a diatomic molecule increases to a certain extent, the two vibrating atoms recede to such a distance

F

that the restoring force is too weak to bring them back. The molecule flies apart, or dissociates. When this happens, the band structure of the absorption spectrum changes to a continuous absorption, and the frequency where this begins gives the energy quantum necessary to dissociate the molecule, i.e. the heat of dissociation.

Another kind of spectrum used is the ultra-violet absorption spectrum. This is most valuable in the case of complicated organic molecules, since specific groups and valency types cause definite absorption bands. The use of such spectra has been of much value in conjunction with synthetic chemical methods in deciding the structures of complicated molecules.

The Electronic Theory of Valency

It has been mentioned that the explanation of chemical affinity as due to electrical forces had been given early in the nineteenth century, but until the discovery of the electron by Sir J. J. Thomson in 1897 little progress could be made in this direction. With the recognition of the electron as a negatively charged particle present in all atoms, it was clear that an electrical explanation of valency was possible. Sir J. J. Thomson and Sir William Ramsay, from about 1900, made many attempts to give such an explanation, but with little success. The chief difficulty was the one which had led to the abandonment of the electrochemical theory of Berzelius, viz. the existence of great numbers of carbon compounds which show no indication of having such electrochemical or polar properties as are found with salts or compounds of charged ions.

When the structure of the atom as a small positive nucleus surrounded by negative electrons equal in number to the nuclear charge had been made out about 1913, some progress could be made in explaining the nature of valency. It was clear that there were two main types of compounds. Some, like salts, consist of separate charged ions, held together by electrical forces, between which there are no true valency bonds, and they are called *electrovalent* compounds. Other molecules, such as ammonia, NH_3, and marsh gas, CH_4, are not ionic but consist of atoms held together by directed valency bonds. These bonds originate in electrical forces, but in quite a different way from the attractions between unlike charges in the ionic compounds. They have fixed directions, as was proved by the facts of stereochemistry, which require, for example, that the carbon atom valencies are arranged tetrahedrally.

A charged ion is formed from a neutral atom when the atom loses or gains a negative electron. If, for example, an electron leaves a sodium atom it will cause the formation of a positive sodium ion. If the electron enters a chlorine atom it forms a negative chlorine ion. The formation of sodium chloride from sodium and chlorine thus consists essentially in the transfer of an electron from the sodium atom to the chlorine atom.

G. N. Lewis, the American physical chemist, suggested in 1916 that in non-polar, or *covalent*, compounds (distinguished from *electrovalent* compounds consisting of charged ions) one electron from each atom contributes to the formation of a pair of electrons shared in common by the two atoms, this shared pair forming the valency bond. If the electrons are represented by dots, the two types of compound formation may be shown as follows:

$$Na . + \overset{..}{\underset{..}{\cdot Cl}}: = Na^+ + \overset{..}{\underset{..}{:Cl}}:^- \quad \text{(electrovalent)}$$

$$H. + \overset{..}{\underset{..}{\cdot Cl}}: = H \overset{..}{\underset{..}{:Cl}}: \quad \text{(covalent)}$$

The chlorine atom has seven electrons in its outer shell, the sodium and hydrogen atoms have one. In the first case the electron leaves the sodium completely to go to the chlorine, and ions are formed. In the second case the hydrogen and chlorine atoms share a pair of electrons and form a valency bond.

The Entry of Wave Mechanics

In recent times a change has come over the theory of atomic structure. The simple picture of an atom as consisting of a small positive nucleus with electrons revolving around it in orbits like planets round the sun, has undergone a transformation in the hands of the new developments of wave mechanics so that no one can now form for himself a picture of an atom—or, if he does, he realizes that it is probably not at all like what is really there. The impact of wave mechanics on the atom has upset many cherished physical and even philosophical ideas, but it has had one surprising result. Whereas the earlier theory of atomic structure proved ineffective in providing an explanation of the covalent bond, this proves relatively easy and simple in the new theory.

The point charge of the electron revolving in an orbit, which from the start offers little hope of explaining a stationary and directed valency bond, now becomes either a cloud of charge spread over a region of space, or a probability of the occurrence of a point electron in that region. For the hydrogen atom, this distribution of charge is spherical and is called an s electron. The two valency electrons of the oxygen atom have quite a different kind of charge distribution; they protrude from the atom volume in two dumb-bell shaped clouds, called p electrons, at right angles. There are two kinds of electrons with opposite spins, and two electrons with opposite spins tend to pair and form a bond. The clouds of the s electrons of two hydrogen atoms approach the ends of the dumb-bell shaped p electrons of the oxygen until the clouds overlap, and thus a directed covalent bond is formed. (The other end of the p dumb-bell remains, since the whole dumb-bell is a single p electron.) Moreover, the theory also gives valency angles. In H_2O the theoretical valency angle is $90°$, but it is actually somewhat larger because of the repulsion of the two hydrogens. The tetrahedral arrangement of the four carbon valencies follows from a more subtle manipulation of blending s and p electrons on the carbon atom to form four hybrid identical tetrahedral bonds.

The requirements of chemistry on physics have proved very exacting. The simple ideas about atomic structure which served to open out an immense field of discovery in atomic physics found no response in chemistry, where the structures of molecules and the nature of valency forces had been known for decades as a result of purely chemical, or physico-chemical, investigations. With the advent of the wave-mechanical theory of the atom the position changed, and physical theory in this field has proved fruitful and suggestive. Although it can hardly be claimed that all the problems about the structure of molecules which the chemist would like answered have been cleared up to his satisfaction, great progress has been made. This is a recent achievement, dating back some ten or so years only.

The Century's Progress

If we look back to the views on molecules held about 1850 by chemists and compare them with those held in 1950, we find that surprisingly little has changed. The chemical atom has lost its glossy impenetrability, but for all the chemist cares in the majority of cases it is still a little billiard ball which can be joined by a bond to another. The size of an atom was known many years ago with surprising approximation to the truth, but the small differences in size which are now known with considerable accuracy are important to the chemist in many ways. If a molecule consists of two parts joined by a bond, and the two parts can rotate around the bond, it is important to know if the atoms in the two halves are small enough to pass one another or not, and this can now be stated.

The shapes of molecules are also important, although the great increase in knowledge of the structure of crystals still fails to give much significant information to the chemist about the way the substances will react. The shapes of isolated molecules, whether rod-shaped, triangular, plane or non-planar, or the like, have in a great many cases been found, as well as the distances between the atomic centres. In the majority of cases these shapes were known to chemists long ago, but the distances were not. These distances are important in another way. The single bond, double bond, and triple bond distances are increasingly shorter in this order. From the bond distance, therefore, the nature of the bond can be inferred. The bond length can be found by X-rays, electron diffraction, or absorption spectra, as explained above. In some cases it is found that the distances are intermediate and the bonds therefore appear to be intermediate between, say, single and double bonds. This is the case with benzene. In such cases the actual bond is a kind of hybrid, and is a result of what has been called *resonance*. The idea that the bonds between carbon atoms in the benzene ring were not alternate single and double bonds, but bonds of a kind intermediate between these and all of them identical, had been proposed long ago by Kekulé.

If we can picture two molecular structures which differ only by the arrangement of electrons in the valency bonds, it is always possible that the actual molecule is not either of these forms, nor the substance a mixture of the two kinds of molecules, but each molecule is in a state intermediate between the two limiting states, or is what is called a resonance hybrid. The theory of valency has become more complicated, but it is still essentially the same theory as that proposed about 1850 by Frankland. This is typical of much of the new knowledge given by physical methods to chemistry; the chemical picture has become more detailed, but its main outlines are still the same. It can cause astonishment to anyone but a chemist to be told that chemists had, by purely chemical methods involving the preparation of different kinds of compounds, and the recognition of isomerism and stereoisomerism, arrived at substantially correct models of chemical molecules long before 1900, but such is in fact the case.

Chemistry is not a branch of physics but a distinct science with its own methods and ideas, and although it can and does make much use of physical methods it is not subordinate to these. It is very probable that if the structures of molecules had not already been fixed as a result of chemical investigation, physical methods would not yet have caught up with them. Modern chemistry makes use of a variety of techniques. It has its own, as in synthetic organic chemistry, stereochemistry, and many branches of physical chemistry. It also uses the results of physics, as in X-ray or electron diffraction methods and spectroscopy. All these different methods give concordant results, and from them the detailed structures of very many molecules are now accurately known.

CHAPTER VI

THE CHEMICAL ELEMENTS

by J. R. PARTINGTON, M.B.E., D.SC.

WE owe the idea of an element to the Greek philosophers, and the four-element theory of Aristotle lasted well into the eighteenth century. The alchemists had postulated three principles (*tria prima*) of salt, sulphur and mercury, but they were not regarded as the definite substances which have these names. It was Robert Boyle in a famous book, the *Sceptical Chymist* (1661), who proposed that chemists should regard as elements those actual substances from which materials can be formed and into which they can be resolved. Lavoisier (1789) adopted practically the same definition, regarding a chemical element as a simple form of matter which cannot by any known means be resolved into simpler forms. Lavoisier drew up a table of the elements then known. It included the imponderable elements heat and light, the gases oxygen, hydrogen, and nitrogen, the non-metals sulphur, phosphorus and carbon, seventeen metals, five earths (such as lime), and the then unknown radicals of three acids. Davy in 1807–8 showed that the earths and alkalis are not elements but oxides of metals (calcium, potassium, sodium, etc.) previously unknown, and in 1810 he proved that what Lavoisier thought was an oxide of a radical was really an element, which he named chlorine.

By 1850, many true elements were known. New non-metals such as selenium, bromine, and iodine had been discovered; heat and light had been removed from the list, and many new metals had been added. In Gmelin's *Handbook of Chemistry* (1849) there are sixty-one elements—twelve non-metals and forty-nine metals—and all these still remain in the list, which has now grown to over ninety elements.

The discovery of a new element was always an outstanding event in chemistry, and as far as was known in 1850 there was no reason to think that the number of elements was limited. The discovery of the spectroscope showed that the ordinary elements are found in other heavenly bodies, but in 1868 a new spectrum line was seen in the examination of the sun, and Sir Norman Lockyer suggested that it belonged to an element not known on the earth, which he called *helium*. We now know that helium is present in the atmosphere of the earth. A new metal, thallium, was discovered by Sir William Crookes using the spectroscope in 1861.

The Classification of the Elements

The old classification of elements into metals and non-metals is not very satisfactory, and many schemes of arrangement have been proposed from time to time. The elements were classified according to valency, but this brought together such unlike elements as chlorine and sodium. The most promising method was based on the atomic weights. As early as 1815–16, William Prout, a London doctor, suggested that all atomic weights are whole numbers, hydrogen being taken as unity, and hence the atoms of the various elements are condensations of hydrogen atoms, which he called the primary matter or *protyle*. Less hypothetical schemes of classification based on atomic weights were proposed. Newlands in 1864 arranged the elements in order of their atomic weights, beginning with hydrogen, and found that like elements appear at corresponding places in the sequence, every eighth element belonging to the same group, and he called this the *law of octaves*. It was not taken seriously, although it contains the germ of the modern method of classification.

In 1869 the Russian chemist Mendeléeff, and in 1870 the German chemist Lothar Meyer, arranged the elements in the order of atomic weights, as Newlands had done, but in a much improved way, some atomic weights in the meantime having been corrected. Mendeléeff found that the elements arranged in this way form *periods* or horizontal rows, each vertical *group* containing elements in corresponding places in different periods, and these groups contain analogous elements. The first group, for example, contains the alkali metals, lithium, sodium, potassium, etc., the seventh group the halogen elements, fluorine, chlorine, bromine and iodine. The eighth group contains what Mendeléeff called transitional elements, iron, cobalt and nickel, and the platinum metals, with atomic weights nearly the same in each group of three. It was his separation of these transitional elements from the rest which enabled him to bring order into the system.

Mendeléeff found it necessary to leave gaps in order to maintain the similarity of properties in groups, and he predicted that these gaps corresponded with elements then unknown. He predicted their properties. His predictions were brilliantly verified by the discovery of the elements gallium (Lecoq de Boisbaudran, 1875), scandium (Nilson, 1879), and germanium (Winkler, 1886). From about 1890 the periodic law was an established feature in chemistry. A modern form of periodic table is shown, in which the elements, represented by their symbols, are in the order of the atomic numbers (see p. 69) given by the figures.

The Rare Earths

From time to time in the nineteenth century new elements belonging to a group called *rare earth metals* were discovered. These are all very much alike in chemical properties and very difficult to separate from each other. One of the prominent workers in the field was Sir William Crookes. The rare earth elements could not be fitted at all easily into the periodic table, since there were too many of them. Crookes, who studied their spectra, also investigated the discharge of electricity in gases at low pressure, and as a result of all this work he reached the conclusion, about 1887, that there are really very many rare earth elements differing only slightly from one another, which he called *meta-elements*. He also revived Prout's hypothesis, suggesting that the cathode rays (see Chapter III) are the same as protyle. An important suggestion made by Crookes was that the deviations from whole numbers of the atomic weights of elements were due to the circumstance that many common elements may be mixtures of different atoms of the element with slightly different weights. Crookes thus anticipated the much later discovery of such varieties of an element, now called *isotopes*. By about 1890, many chemists believed that all the elements are derived from one or a few kinds of primary matter, and the most promising primary matter was hydrogen. This view is now known to be substantially correct (see Chapter IX).

One difficulty about the rare earths was that it was not known

PERIODIC TABLE

A Groups							Transitional			B Groups							Zero Group
I	II	III	IV	V	VI	VII	VIII	VIII	VIII	I	II	III	IV	V	VI	VII	0
H 1																	
Li 3	Be 4											B 5	C 6	N 7	O 8	F 9	He 2
Na 11	Mg 12											Al 13	Si 14	P 15	S 16	Cl 17	Ne 10
K 19	Ca 20	Sc 21	Ti 22	V 23	Cr 24	Mn 25	Fe 26	Co 27	Ni 28	Cu 29	Zn 30	Ga 31	Ge 32	As 33	Se 34	Br 35	A 18
Rb 37	Sr 38	Y 39	Zr 40	Nb 41	Mo 42	Tc 43	Ru 44	Rh 45	Pd 46	Ag 47	Cd 48	In 49	Sn 50	Sb 51	Te 52	I 53	Kr 36
Cs 55	Ba 56	57–71 Rare Earths	Hf 72	Ta 73	W 74	Re 75	Os 76	Ir 77	Pt 78	Au 79	Hg 80	Tl 81	Pb 82	Bi 83	Po 84	At 85	Xe 54 / Rn 86
Fr 87	Ra 88	Ac 89 / Trans-uranic 93–98	Th 90	Pa 91	U 92												

La 57	Ce 58	Pr 59	Nd 60	Pm 61	Sm 62	Eu 63	Gd 64	Tb 65	Dy 66	Ho 67	Er 68	Tm 69	Yb 70	Lu 71

Np 93	Pu 94	Am 95	Cm 96	Bk 97	Cf 98

how many there were. New ones were discovered from time to time, but some old ones were shown not to be individuals but mixtures of other earths. It was only after the discovery by Moseley (1913-14) of the X-ray spectra of elements and the proof that a unique number for an element, roughly its number in the sequence of atomic weights, starting with hydrogen, could be found from such spectra (see Chapter IV), that a roll-call of the elements could be made. Owing to the fact that some elements are mixtures of isotopes, the order of *atomic numbers*, which Moseley identified with the positive charges on the atomic nuclei, sometimes differs from that of the atomic weights, and the latter are not so significant. In the modern periodic table, the elements are arranged in the order of their atomic numbers.

Moseley found that there should be fifteen rare earth elements between lanthanum and lutetium, including one as yet unknown, and this solved the problem of the number of such elements. It was still difficult to place them in the periodic table. This was first satisfactorily done on the basis of a knowledge of the structures of the atoms by Bohr. He showed that in passing from one rare earth element to the next, the positive charge on the nucleus increases by one unit, and one electron is added to the shell of the atom. This electron, however, does not go into an outer, or valency, shell, but deep inside the atom in an incomplete shell, so that the valency shell in the outer part of the atom remains unchanged and the properties of all the rare earths are very much alike. When this inner incomplete electron group is filled, the rare earths come to an end, and the next element, hafnium, belongs to a different group.

The Inert Gases

In 1894 the scientific world was startled by the announcement by Lord Rayleigh and Professor (afterwards Sir William) Ramsay that the atmosphere, which had been analysed repeatedly by the most expert chemists, contains no less than one per cent of a previously unknown gas which had always been obtained mixed with atmospheric nitrogen. Rayleigh's careful measurements of the density of nitrogen had shown that atmospheric nitrogen is always a little heavier than nitrogen obtained from chemical compounds, and this had led him to suspect that a heavier gas was present in the former. The new gas had surprising properties; it forms no compounds with any other element and its molecules consist of single atoms, unlike those of most common gaseous elements, which are diatomic (O_2, N_2, H_2, etc.). It was called *argon*, from a Greek word meaning lazy or indifferent. In seeking a place in the periodic table for argon, Ramsay concluded that it must belong to a new group, and hence there should be other unknown inert gases in this group. He soon found a second member in helium, which had been discovered in the spectrum of the sun in 1868. Other inert gases in the atmosphere, neon, krypton, and xenon, followed, and the radioactive emanation, radon, was later recognized as the heaviest member of the inert gas group. Since all its members have zero valency, Ramsay called it the zero group, but it really closes the relevant periods, the atoms building up external electron groups to a number of eight in an inert gas (see Chapter IV).

Structure of the Periodic Table

It had been noticed that the numbers of elements in the periods of the periodic table, beginning with the one containing hydrogen and helium, are 2, 8, 18 and 32, the last period, which contains radio-active elements, being a fragment of what is probably a longer period. The explanation came with the theory of atomic structure, according to which the outer electrons of the atom, added as the nuclear positive charge increases step by step from the lightest to the heaviest atom, are arranged in successive shells. Every electron is characterized by four so-called quantum numbers, which specify its energy, and according to a rule proposed by Pauli (1925) no two electrons can have all four quantum numbers the same. This leads at once to the observed numbers of elements in the periods. Even if the actual elements were unknown, their places could be predicted and the whole periodic table built up on this theoretical basis.

Isotopes

For Dalton a chemical atom was incapable of change in any way, and all the atoms of the same element were identical. In 1913 it was known that the radio-active elements uranium and thorium break down spontaneously with emission of alpha-particles and beta-rays, giving a sucession of products. The last product is an element which, from the change of mass of the initial atom, calculated from the number of alpha-particles emitted, should have an atomic weight of 206 or 208, very nearly the atomic weight of lead, 207. The two kinds of lead, uranium lead (atomic weight 206) and thorium lead (atomic weight 208) are two varieties of the element lead, which were called isotopes. Later work by J. J. Thomson and F. W. Aston showed that many common elements are mixtures of isotopic varieties, the observed atomic weight being an average value. Common lead contains the isotopes of atomic weights 206 and 208. The element chlorine, of atomic weight 35.5, is a mixture of two isotopes, 35 and 37; oxygen contains the isotopes 16, 17 and 18; carbon the isotopes 12 and 13; and so on.

Artificial Elements

Although radio-active changes lead to the formation of lighter elements by the spontaneous breaking up of atoms, the reverse process of building up heavier atoms from lighter does not seem to occur naturally on the earth, although it may do so in the sun. Rutherford (1919) showed that swift alpha-particles can penetrate to the nucleus of the nitrogen atom and eject hydrogen nuclei (protons) from it. Presumably the alpha-particle enters the nucleus but this is unable to digest it, and the extended nucleus breaks down. F. Joliot and Mme. I. Curie (1934) then found cases where the captured particle can stay in the nucleus for an appreciable time, and so a heavier atom can be synthesized. Since then enormous numbers of artificial elements have been obtained by the capture by atomic nuclei of swift protons, deuterons (heavy hydrogen nuclei), alpha-particles (accelerated helium nuclei) or neutrons.

The most spectacular achievement in this field is the production of elements heavier than the heaviest known natural element, uranium (atomic weight 238). These so-called *transuranic* elements, named *neptunium, plutonium, americium, curium, berkelium* and *californium,* follow uranium in succession in the periodic table, and are all radio-active. It seems very probable that this last period in the table will be enriched by other transuranic elements.

Many of the artificial radio-active elements break down by the emission of positive electrons, which are not emitted by natural radio-active elements. Each has a characteristic half-life, i.e. its activity decays to half a given value in a definite time, and hence the element can be characterized by its radio-active properties. Since these elements follow the common elements through series of chemical changes, they can be used as so-called *tracer elements*, and valuable information about many kinds of chemical reactions has been obtained by their use.

By using the radio-active carbon atom as a tracer element, carbon metabolism can be followed. One of the most interesting results in this field is the discovery that the oxygen given off in the photosynthetic reaction in green plants under the influence of light does not come from the carbon dioxide absorbed by the plant, as was always thought, but from water. In such work it is not necessary to use radio-active tracer elements, since the heavier non-radio-active isotopes of elements such as carbon, hydrogen and oxygen can be used, but since radio-active measurements are much easier than those with non-radio-active isotopes, many more workers have been attracted to this field since they became available. One of the objects of the construction of an atomic-energy pile, such as that at Harwell, is to provide radio-active tracer elements. They may also have uses in medicine.

Allotropy

It was recognized early in the nineteenth century that some elements can exist in different forms, and this phenomenon was called *allotropy* by Berzelius (1841). Sulphur exists in at least two different crystalline forms and several non-crystalline or amorphous varieties. Phosphorus is known in the active white form (commonly mis-named yellow phosphorus) and in the much less active red form. Oxygen is known in the ordinary form of oxygen gas (O_2) and in the form of ozone (O_3), the latter having a powerful smell and very active oxidizing properties not possessed by ordinary oxygen. The most spectacular example of allotropy, however, is that of the element carbon. This occurs in the three forms of charcoal, diamond, and graphite (or black-lead). Chemists had convinced themselves by the end of the eighteenth century that diamond is merely pure carbon, by showing that equal weights of charcoal and diamond, when burnt in oxygen, gave equal weights of the same gas, carbon dioxide. There was some doubt as to whether graphite was a form of carbon or a compound of carbon and iron, since all specimens of native graphite leave some oxide of iron on combustion. The fact that pure graphite is a form of carbon was proved by Brodie (1855).

Since they all contain the same element (carbon), charcoal, graphite and diamond should be convertible into one another. Foucault and Fizeau (1844) converted charcoal into graphite by heating it to a very high temperature in the electric arc, and much artificial graphite is now made for use as a lubricant by essentially the same method. The conversion of charcoal into diamond is much less simple. Many attempts to carry out this change were made in the nineteenth century, and in 1893 the French chemist Moissan (who had discovered the very active fluorine in 1886) announced that he had obtained artificial diamonds. He made use of the electric furnace, which he had perfected shortly before, to heat iron and carbon to an intense white heat. The crucible was then plunged into cold water, and when the iron was dissolved by an acid, microscopic diamonds, both black and transparent, were found in the residue. Moissan proved that these were diamonds by burning them in oxygen and in other ways. He tried very many other methods of preparing diamonds, without success, and he emphasized that unless the details of his process were carefully followed, no diamonds would be obtained. It is, therefore, not surprising that many subsequent workers who have used different processes (many of them already tried by Moissan) should have failed to obtain diamonds. Ruff, in 1917, by carefully following Moissan's procedure, obtained diamonds, which he characterized in several ways.

Another interesting allotropic form of an element is the so-called active nitrogen, most of the work on which was done by the late Lord Rayleigh (1875–1947), who was its real discoverer (1911). It is form.d by the action of an electric discharge on nitrogen gas at low pressure. Unlike ordinary or molecular nitrogen (N_2), which is of small activity, combining only at high temperatures with other elements, active nitrogen combines readily with many elements and enters easily into reaction with other substances. Active nitrogen may consist mainly of nitrogen atoms but its real nature is still in doubt.

The Definition of an Element

A definition is stated in a dictionary to be a brief description of a thing by its properties and it is so understood in practical science. It is a nice point as to when a definition contains so many words or full-stops that it becomes a description. Practical science agrees with daily life in setting small store in definitions; it is no easier to define an element than to define a dream, a common cold, or a dog. Sir Arthur Eddington once defined physics as the science described in any large handbook of physics, and the definition of an element as a substance named in a list of elements in a book on chemistry is no worse than this. The names have got into this list as a result of an enormous amount of practical work, the description of which fills thousands of pages of journals and books, and it would indeed be a miracle if any brief collection of words could summarize this information. At the present day there are also hundreds of artificial elements. Most of these are radio-active isotopes of common elements, but three of them, with atomic numbers 43, 85 and 87, fill places in the periodic table which were previously empty. All the isotopes of each of the other elements are counted as merely varieties of the same element.

It is very difficult at the present time to give a satisfactory definition of a chemical element. It was once said that an element is a form of matter all the atoms of which are identical. The existence of isotopes negatives this definition. Another definition was that an element is characterized by the positive charge on the atomic nucleus (the so-called atomic number), but there are different radio-active isotopes (e.g. uranium Z and uranium X_2) with the same atomic number and atomic weight, and isotopes of different elments with the same atomic weight. The proposal to regard all the isotopes of an element as varieties of the same element goes back to the days when isotopes of radio-active elements, or elements formed by their disintegration, were supposed to be chemically inseparable, and is inadmissible in face of the facts that hydrogen and deuterium have different properties and are easily separated; the separation of isotopes of other common elements such as oxygen is now a common operation. The chemist's definition of an element as a form of matter which cannot yield, in a complete chemical change, any different form of matter of smaller weight (often paraphrased by the layman as a form of matter which cannot be split into simpler forms, although chemists do not go about their business with hatchets) runs into difficulties when it is known that many kinds of matter regarded as elements can be resolved into parts by the action of radiations of high energy, such as the alpha-particles, which eject hydrogen nuclei from nitrogen atoms. It is still true that such changes do not occur in the kinds of change in which chemists are primarily interested, which are brought about in the test-tube or in the furnace, and it is on the results of such changes that the table of elements is constructed.

The energies called into play in nuclear chemistry are enormously greater than those intervening in the chemical laboratory and belong to an entirely different category. Ostwald (1904) foreshadowed this situation and related the existence of chemical elements with the magnitudes of energy changes. The present description is not very satisfactory, but it is the best we can give.

Chemists had long suspected that the chemical elements were not really the simplest forms of matter. Prout's suggestion that hydrogen is a constituent of other atoms, and Davy's (1812) that electricity enters into the composition of atoms, have both been verified. The periodic table was arrived at as a result of chemical researches, and it gave the clue to the structure of the atoms of the elements. The main change which has come over the views on the nature of the elements held since 1850 is that their atoms are now regarded as composite particles, and the genetic relations between elements which was suggested by the periodic table is now explained in terms of atomic structure. In practical chemistry the elements still remain simple forms of matter and the table of the elements still has validity. Only in very exceptional circumstances do the elements cease to be what they were until a knowledge of their structure was achieved.

CHAPTER VII

GEOLOGY

by W. T. GORDON, M.A., D.SC., F.R.S.E.

IT is scarcely an overstatement to say that the generation that saw the opening of the Great Exhibition of 1851 was the witness of the most stupendous change in the geological philosophy of mankind. By that date the basic theories of geological philosophy had been enunciated; but their acceptance by the general body of geologists, much less by the general body of the public, was not whole-hearted. As a fact, like most theories, the new views were often stated in over-simplified terms, and, in the days that have supervened, they have undergone modification.

But to secure any appreciation of the public's reaction, in the middle of last century, to such notions, it must be remembered that geological ideas have very special relations to all human philosophies. No science touches humanity more intimately, for every human being is a part of nature, and strives to adjust himself satisfactorily to his surroundings, both physically and mentally; and geology, being the history of the Earth and its inhabitants, must, of necessity, figure largely in securing that adjustment.

The Uniformitarian View

So long as large areas of the earth were virtually unknown, efforts to attune philosophies to Nature resulted in changing and often fantastic conclusions. Although the friar Generelli in 1749 and Desmarest in 1777 devised systems that did not demand extravagant hypotheses, and although James Hutton in 1785 came nearer to the Uniformitarian view than any other, yet these attempts to establish fixed principles in geology were universally held in disrepute at the beginning of the nineteenth century, despite Playfair's eloquent and skilled defence of the last-mentioned author.

But the accumulation of geological observations had rendered suspect the older notions, and had left the Uniformitarian doctrine of Lyell, published in 1829, the only feasible one on which to base a philosophy of Nature; namely that the present and the past phases of the Earth's history could be interpreted in terms of processes *now* observable in Nature. But—and it was a very substantial but—older schools of thought, the Neptunist and the Plutonist, had still their protagonists, especially as the catastrophic implications, inherent in both these philosophies, appeared to offer more adequate explanations for certain events described in the Bible, and consequently received some kind of theological sanction. While, therefore, those best qualified to have an opinion rejected the older philosophies, the new doctrine was none-the-less combated fiercely at every stage, and geologists were themselves the most relentless critics, perhaps because they alone saw the full consequences of acceptance.

Neptunism and Plutonism

Now the older notions had not been based on imaginary phenomena. The Neptunists could point to enormous accumulations of sediments formed in water or by wind. (Ice had not then been noted as a possible vehicle for the formation of sediments.) The Plutonists or Volcanists were equally able to show stupendous volumes of rock undoubtedly the result of volcanic action. In both cases, too, the amounts of such rocks were far greater than could be accounted for by the activity of observable processes in action. The epithets 'Neptunism' and 'Plutonism' were applied to the extreme statements of the respective philosophies and neither denied the existence of the process advocated by the other, but only its dominance.

The over-ruling factor in the acceptance of any of the theories was the question of Time. Desmarest had shown in 1777 that the river valleys in the Auvergne *might* have been carved out by the rivers then occupying them, but that it would have required a long time. The Uniformitarian view required long periods of action for present-day processes, if the huge thicknesses of rock involved were to be explained in terms of these processes, even on the basis of the knowledge then available.

Scriptural Chronology

Now, all the philosophical theories of the day accepted the Mosaic account of the Creation and assigned a very short period to the existence of the Earth, and Uniformitarian views seemed *ipso facto* contrary to scriptural doctrine. Yet as far back as 1804 Dr. Thomas Chalmers, the Scottish divine, lecturing at St. Andrews, stated that "the writings of Moses do not fix the antiquity of the globe". This courageous position, taken after weighing the evidence accumulated by the geologists, showed his sagacity. Later (1814) he thought that vast ages might have intervened between the Creation of the earth and the first day recorded in Genesis, but he held the successive days of the Creation, as given in that book, to be natural days of twenty-four hours. Between that date and the opening of the Great Exhibition every step in advance was fiercely contested, but by 1850 the whole picture had changed.

The accumulation of geological data, the wranglings of philologists as to the exact meanings of the words in Genesis, the bitter quarrels between the exponents of this, that or the other aspect of philosophy, had somewhat subsided, and, as indicated by Hugh Miller in his *Testimony of the Rocks* (1854), the consensus of opinion seemed to be that the Earth was of unknown, but probably considerable, antiquity; that the days of the Creation as given in Genesis were long periods, and not days of twenty-four hours, and that man—the last of the creatures formed—dated back to 4004 B.C. This last date was printed in our Bibles even in this present century.

Of course, there were, and for that matter still are, many who considered these opinions false; but, at the date of the opening of the Exhibition, this represented the philosophy of the average educated man.

So far, only processes and chronology have been mentioned in this digest, for these alone figured in the build-up of geological philosophy up to 1851; but, in every branch of the study, data were being accumulated. Minerals had been elaborately described, and their characters tabulated; the ordinary microscope had been invoked to assist these studies, and the reactions of plates (thin slabs) of minerals in plane-polarized light had been noted; fossil animals and plants had had similar attention paid to them, and their uses as stratigraphical indices had been recorded. Indeed the application of the microscope to fossil plants had set a fashion—the first records of which date back to Hooke's *Micrographia* (1665)—of investigating the minute cellular structure of these fossils. In other words the ultimate structural elements of certain fossil types had become a commonplace topic in investigation.

But with all this accumulation of data there was little philosophical advance: something was wanted to give a soul to these dry details. Lamarck's work in the biological field had been sadly neglected. In 1838 Darwin records that "after my return to England it appeared to me that by following the example of Lyell in Geology, and by collecting all facts which bore in any way on the variation of animals and plants under domestication and nature, some light might perhaps be thrown on the whole subject", i.e. the subject of evolution in the organic world.

Early Evolutionary Theory

Robert Chambers's *Vestiges of Creation* (1844) went into twelve editions; and, while he held evolutionary views, he considered that organisms arose as conditions favoured them and gradually developed from the simplest types. Darwin, in the same year, had actually written the first draft of his *Origin of Species by Natural Selection*, and had arranged for its publication, in the event of his death before he could revise it. (In the end it was not published until 1859.) Lyell and his contemporaries, while recognizing that Uniformitarianism in the inorganic world had its counterpart in the organic world, were not ready to propound, and defend, an evolutionary process in that organic world. The real stumbling-block was the want of some *process* of evolution, and this was not forthcoming.

Thus, around 1851, a satisfactory geological philosophy in the inorganic world had been attained and established on a solid foundation of observation, so far as the criteria of the times would allow. In the organic field the bomb-shell had not yet burst, but indeed the delayed fuse was already set and only wanted the trigger action of Wallace's letter to Darwin to explode it. This did not come until 1859, however, and the dawn of an entirely new philosophy of life was not yet in sight when the Exhibition opened.

On the material side of geology the Exhibition showed how raw materials were being provided for many industries. Geological evidence was proving the presence of useful minerals and rocks, often concealed under other strata. But these benefits to the trade of the country, though the undisguised *raison d'être* of the Exhibition, do not reflect the state of geological theory and inquiry at the time so remarkably as the more philosophical aspects of the subject.

The Approach to 1951

And now what of the approach to 1951? Like all great movements, the emancipation from the thraldom of supposed authority went too far. The yardsticks that had sufficed to free geological thought have been proved, in many cases, to be inaccurate, but the inaccuracy has shown that the earlier conclusions were conservative and in no cases have the trends of those conclusions been proved false.

Adopting then a parallel course to that employed in assessing the position around 1851, what are the present ideas of geological chronology?

Geological Chronology

Although the uniformitarian doctrine had been established, so far as inorganic nature was concerned, and Darwin's *Origin of Species* had extended its scope into the organic world, and although the principle of Steno, the Dane, that the topmost rocks in an undisturbed succession were the latest to be formed, was generally admitted, yet there was no international agreement as to the nomenclatures of the suites themselves. Lehmann, William Smith and others made classifications that were employed, both in this country and abroad; but correlation with suites at considerable distances was either not attempted or was rather unsatisfactory. Even the circumstances that might render Steno's law of superposition inoperative in definite areas were not fully appreciated.

In the British succession a table of strata had become established consequent upon the works of William Smith, Murchison, Sedgwick and Lyell, and relatively few alterations to it have been made in the succeeding years. Smith employed local names—mainly those used by quarrymen—for the formations that stretched diagonally across England from Dorset to Yorkshire and that could be more or less identified with Lehmann's Secondary rocks.

Murchison and Sedgwick strove to unravel the sequence of strata beneath those studied by Smith, and Lyell specialized in differentiating the strata that succeeded them, but employed names based on more scientific criteria. While the broad outlines of this classification were accepted, after others had been tried out and found unacceptable to the majority of geologists, a regional character was reflected in the naming of smaller assemblages of strata. Thus Mountain Limestone Formation indicated the character of the rocks of the Pennine Chain, Bath Oolite those of the strata around Bath, and Cretaceous rocks those associated with the chalk of N.-W. Europe (Lat. *creta*=chalk). On a more scientific principle, Lyell differentiated the rocks overlying those of Cretaceous age by estimating the proportion of modern shells found in them, and the whole set were therefore called Cainozoic (Gk. *Kainos*=recent). Thus the groups *Pleistocene*, *Pliocene*, *Miocene*, *Oligocene* and *Eocene* were distinguished, their names being derived from Greek words indicating the transition from the most recent to the most remote, and all in relation to the proportion of the species of recent shells found in them.

In other lands other names were applied—*Dias* and *Trias* in Germany for the rocks that could be classified into two or three suites respectively: and similarly in other countries. Again, more local names or characteristics were applied to indicate smaller divisions such as *Muschelkalk* or *Kupferschiffer* or *Autunian*, etc.

Nomenclature

But, as a result of discussions at successive International Geological Congresses and usage by research workers in the past century, the British general scheme has been universally adopted as a basis, and other names employed, in preference, only in a few cases. Modifications have been made, and some others may be made in the future, but the succession as determined in Britain has proved the most acceptable. Of course, it may seem somewhat inappropriate to call certain rocks in New Zealand Cretaceous, when no chalk occurs among them but, on the contrary, beds of excellent coal. The name Carboniferous might have been more apposite. The names therefore must be taken as tokens, not as accurate descriptions.

Yet, aside from mere priority in usage, there is a deal of justification for adopting the names established in Britain. In no other land can so many different suites of rock be studied in so small an area. Aside from those of Miocene age, examples of practically every large formation are found in Britain; and, indeed, round Bristol, rocks of nearly every geological age may be seen within a range of one day's excursion. Little wonder then that stratigraphical geology made such rapid strides in Britain, or that the names applied by British geologists have received international sanction. None-the-less, in a few cases the British sequence has become recognized as of too specialized a type, or of too restricted an extent, to be universally accepted. Thus *Jurassic System* has been given preference to *Oolitic System*, though the latter name may have survived for a smaller suite.

As we approach 1951 then, the following table is the touchstone by which the rocks of the earth are classed in a broad sense. The table also contains estimates of the thickness of the particular system of rocks, and of the age, in years, as determined by recent calculations. But it must not be forgotten that each of these systems contains rocks of the same kinds—sandstones, limestones, clay rocks and perhaps igneous rocks; it is the period of formation, and not the type of rock formed, that is emphasized.

General table of the stratified surface rocks of the Earth and their thickness and age in years, according to present knowledge

Main group name (Era)	System	Thickness in feet	Age to beginning of each System (two estimates) in millions of years	
Quaternary	Holocene (Gk. holos, complete)			
	Pleistocene (Gk. pleiston, most)	4,000	2	1
Tertiary or Cainozoic	Pliocene (Gk. pleion, more)	18,000	—	15
(Gk. kainos, recent and	Miocene (Gk. meion, less)	21,000	—	35
zoon, a living creature)	Oligocene (Gk. oligos, few)	15,000	—	45
	Eocene (Gk. eos, aurora)	23,000	60	70
Mesozoic	Cretacous (Lat. creta, chalk)	64,000	125	140
(Gk.mesos, middle and	Jurassic (Fr. Jura, Alps)	22,000	157	170
zoon, a living	Triassic (Ger. Trias)	25,000	185	195
creature.)	Permian or New Red Sand-			
	stone (Perm in U.S.S.R.)	18,000	223	220
	Carboniferous	40,000	309	275
	Devonian and Old Red			
Palaeozoic	Sandstone	37,000	354	320
(Gk. palaios, ancient	Silurian	20,000	381	350
and zoon, a	Ordovician	40,000	448	420
living creature.)	Cambrian	40,000	553	520
	Torridonian ⎤	Uncertain	1750	2000
Azoic (without life types)	Dalradian ⎢ British		A late estimate gives	
	Moinian ⎢ sequence		2700 to 3000 since	
	Hebridean ⎦		the earth was first formed	

Now in applying this yardstick to classifying the sedimentary rocks of the earth, the main principle is that the strata of any particular age contain the same suite of fossil remains. Sometimes other, and less certain, criteria are employed, such as lithological characters (Gk. *lithos*, a stone), or even the general colour of the rocks (Old Red Sandstone). Naturally, when no fossil contents occur in a rock suite, these are the only tests available, and mistakes may easily arise. For instance, the Torridonian sandstone was long thought to be of Old Red Sandstone age; but when Cambrian sandstones were found above these beds the true age of the rocks was established. Further, it must be remembered that in no area does a continuous succession of strata exist; there are gaps in the succession and the above table has been built up by correlating partial successions in separated areas. Now marine series are, generally, more continuous than others, and therefore have been accepted as normal, while estuarine, fresh water or land accumulations are not so regarded. Again, volcanic sequences, though many thousands of feet thick, cannot be regarded as more than local in character, though an occasional layer may sometimes contain evidence on which they may be compared with the standard marine succession. Yet rocks of an igneous character that were injected in a molten condition among sediments have recently supplied evidence by which the years that have elapsed since their consolidation can be ascertained, as we shall see. Attempts to determine actual ages of sediments in years by direct means have also been made, and the whole problem is one that occupies attention at the present time.

Methods of Dating

If the same type of sediment always took the same time to accumulate, and we could estimate that rate, then, provided a continuous sequence could be discovered or even a series of discontinuous suites could be correlated with one another, it might be possible to arrive at reasonable figures. In certain lake deposits of late glacial date, alternating layers of coarse and fine materials have been found, and by observation of the formation of these layers another yardstick has been discovered. When the lakes are thawed and streams carry sediment into them the material of a fine nature is held in suspension while the coarser material sinks to form a layer on the bottom. When winter comes and the lakes and streams are frozen, the water beneath the crust is quiet and the finer materials are deposited. This process continues season after season and a succession of double layers (*varves*) accumulates. In an *open* year the varve would be thicker, and the succession of such years is not regular, so a template could be constructed from such layers as 'markers' to compare and correlate the deposits of different lakes; for an open year would give a similar thick layer in many adjacent lakes. At the same time the actual number of varves could be counted, and in this way the actual age in years of quite thick deposits has been determined.

Again the rate of accumulation of limestones has been determined in recent oozes and by other methods and has been employed to ascertain the time taken for beds of such rocks to form. But both methods are subject to considerable possible errors and other methods are equally fallible, so that no great confidence can be placed on the figures that have been obtained, even when maximum thicknesses have been used for the successions and for the amount formed in any given time. But ideas of the extent of geological time as a whole have been computed by methods allied to these.

For instance the common salt in the oceans of the earth has been derived from the continual scouring of the lands by rain and has been carried down to the oceans by rivers; furthermore, salt is not removed from the ocean by organic life, nor has the solution ever been so concentrated that the salt became deposited over the whole ocean floor. Therefore a calculation of the amount of salt in the ocean, and division of the figure obtained by that of the salt brought down by streams each year, should provide a figure giving the age of the oceans. Of course, there are conditions where salt may be removed from the ocean by evaporation, and others where the salt carried in streams never reaches the ocean (areas of inland drainage like the Great Salt Lake of Utah or the Caspian Sea), so the resulting figure for the age of the oceans is much lower than can be accepted. None-the-less, in the early years of this century—when, remember, the average 'man in the street' still held the idea that the world came into being somewhere about 4004 B.C.—these calculations seemed to establish a new order of thought about the duration of geological time, inaccurate though it could be shown to be.

The Radio-active Index

In the last half century, however, an entirely different approach to the problem has been found in the rate of disintegration of radio-active substances in rocks. Minerals containing these substances have crystallized out of certain igneous rocks, notably granites, and the relative ages of many granites have been determined and compared with the sedimentary rocks given in the foregoing table. Now ·the radio-active substances decay, as time goes on, at what we believe to be a regular rate. In the process the gas helium is given off, and radium-lead is left behind. This lead can be distinguished from ordinary lead (see Chapter VI), and hence the establishment of the lead/helium ratio for the particular radio-active mineral gives a means of determining the length of time it has been disintegrating— that is, the date in years when that particular rock was consolidated can be ascertained. Many cases have been examined and the figures obtained have shown a good deal of consistency.

Now granites have been injected into other rocks at several geological periods, and so the date of the beginning of each of the geological systems has been obtained. In the table (p. 104) two separate calculations are given and the general consistency can be seen at a glance. Possible sources of error can be indicated, no doubt; but, unless more accurate figures can be obtained in the future, the present position can be considered satisfactory.

Rock Movements and Changes

But rocks may have suffered many vicissitudes since their formation. Some have been bent into great folds by earth movements, some have been strained beyond their elastic limits and have parted, while their broken ends have moved during readjustment to the extent, maybe, of thousands of feet. In other cases the strains have been taken up by great internal changes. While some phenomena were recognized under the head of 'troubles' in mining operations, the acceptance of folds and 'fault' movements involving great areas, or the actual lateral transport of continents, is of more recent date, though now universally recognized. And under the title of *meta-*

morphism the internal changes of rock minerals effected by pressure and heat are constantly being studied.

Problems associated with many of these movements are frequently of great economic importance as may be seen in our coal and oil fields, our ore deposits and our potential or actual water supplies from underground sources; but the nature of the forces that caused the movements is often a matter of controversy. The geologist can show that the simplest explanation would involve continental displacements, but the geophysicist and astrophysicist cannot always account for the enormous forces that would be required to effect these changes. Lately, magnetic variations in the earth have been attributed to flow movements near the core, and it is not impossible that such flows may be communicated to higher levels in the crust. The geological observations are quite clear, and they have opened up a field for speculation as to their explanation that the workers of last century never contemplated. A completely satisfactory solution is not yet available, however.

Geology and Economics

Turning to the field of economic geology, how can the influence of the discovery and exploitation of coal and oil fields on our modern life be adequately assessed? From the earliest times bitumen, oil and coal have been recorded and the first two, at any rate, exploited for some purposes, though coal was not much used. But the inventions of the steam-engine, and, in these latter days, the internal combustion engine, have revolutionized our life. And the exploration for, and exploitation of, all types of fuel for these and other purposes has meant a scouring of every land for surface indications of underground supplies, while drilling to depths of nearly two miles has been undertaken in suitable places. More economic employment of these fuels is constantly being sought, and, as we all know to our discomfort, rationing had to be imposed upon us with regard to their use. But suggestions for the exploitation of resources that cannot at present be tapped, such as the underground distillation of thin coal seams, may bring some relief in the future.

Then again we are on the threshold of an entirely new source of energy—atomic energy. Visible sources of this supply are elements of high atomic weight, and these are rare in concentrated form. Yet geologists in every land are investigating possible resources at this very moment, and time and again new discoveries of these elements are being recorded.

Geological research also attends the solution of one of the most pressing household problems of city life—the provision of an adequate water supply. There are places where only the water contained in the interstices of rocks is available, and a few where the supply comes from the water trickling into the galleries of mines which has to be pumped up and purified before use. Even steam jets from volcanic orifices have been condensed for this purpose.

The development of instruments delicate enough to detect minute differential effects of electric fields or gravitational fields, or the reflection of explosion waves or 'wireless' waves from materials and structures in rocks hidden under the earth's surface by superincumbent accumulations, has greatly assisted in detecting oil supplies, water supplies and mineral veins in places where they were only half suspected. The 'wild-cat' activities of former days are being largely controlled by magnetic, gravitational or reflectional surveys to supplement the surface geological surveys of the areas in question.

Yet again the provision of building materials is an aspect of geology involving much research. Stones of all kinds, natural or artificial, bricks, cements and plasters present problems in their use that the geologist can assist in solving. The stabilization of soils for the runways of airfields, and the provision of road materials, require the attention of the geologist as well as the engineer. Under the heading of *thixography*, the problems of clays are being actively explored.

Agriculture, too, is demanding increasing quantities of limestone, rock phosphates, gypsum and certain soluble substances like sodium nitrate to stimulate the soil to yield increasing crops, and geological problems attend the provision of all such natural products.

Attention has been directed to the provision of materials for our everyday life, but, once these are used, it may be necessary to dispose of waste materials. To provide a large water supply involves provision for a large sewage disposal plant. Many cases of epidemics of disease can be traced to inadequate or defective sewage schemes. The siting of such plants may involve geological factors, and the planning of new towns must take these problems into consideration. Cities and settlements have had to be abandoned in the past, when either the supply or the disposal of water has not been adequately thought out.

While these aspects of geology are evident, and are clearly in operation at the time of the Festival of Britain, there are others, in the field of pure science, that may not be so obvious and yet may hold great hopes for the future.

Mineralogy

Many minerals have little or no use at present, but they constitute sometimes the bulk and sometimes a minor part of rocks. Their determination in coarse-grained rocks may be easy, but in fine-grained rocks this is not so, and the microscope is now an essential part of the equipment of the geologist. In the examination of minerals the microscope has long been in use, but even in 1867 David Forbes stated that aside from Sorby's paper (1858) there was 'literally nothing' published, to guide the geologist wishing to study with a microscope. How different is the situation today! Modern microscopes permit the geologist to examine, describe and determine the ingredients of rocks with greater and greater precision, and without having recourse to chemical analysis, which, of course, destroys the materials examined. Armed with the data thus acquired, he contrives to reconstruct the past history of the particular rock back to the original conditions of its formation, and to discuss the probable causes of the stages in its history. Moreover the material remains intact, as a rule, and the observations may be checked in the future. The petrological microscope, employing both ordinary and plane-polarized light, has helped enormously in the task, but the reflexion microscope, the ultra-violet light microscope, the phase contrast microscope, the fluorescence microscope, the X-ray spectrometer and the electron microscope are now available, and the future may see great developments in their use in the study of rocks. Of course, the X-ray spectrometer has already been applied most conspicuously in recent years in the realm of crystal atomic structure. All have their limitations; but none-the-less, the plaint of Forbes has lost its sting, and the petrologist of the future has a plethora of instruments, and instruction in their uses, at his disposal.

Palaeontology

While the mineralogist and physicist have been seeking the atomic structure of minerals, and thus attaining a knowledge of the fundamental structure of matter, and while they enter the present period equipped to probe still further into the secrets of nature in the world of crystallography and mineralogy, palaeontological research has been attempting similar investigations in the organic life of the past. Two main lines of development occur; one as an ancillary to stratigraphical geology, and the other an investigation of the biological and environmental conditions of plants and animals in former ages.

So long as fossils, whether plant or animal, were used as mere stratigraphical indices (and to many geologists that is still their only significance) there was no science of palaeontology. Much detailed collection, description and collation was done, but the specimens could all be regarded as 'special creations' without much difficulty, though most thinkers were persuaded that somehow or other the organisms of one horizon had been derived from those whose remains occurred on a lower horizon. With the publication of Darwin's and Wallace's ideas of evolution by natural selection in 1859 a new outlook on palaeontological work was initiated. The outcome of much speculation, experimentation (such as Mendel achieved) and correlation has produced the science of heredity—today practically a separate science; but evolution, by whatever means it was effected, is a working hypothesis so far as palaeontology is concerned, and one that is exceedingly useful so long as its limitations are recognized.

No breeding experiments are possible, and what the palaeontologist calls a *genus* or a *species* is based on form alone, and the selection of what are to be regarded as the essential characters of that form reflects the personal opinion of the worker. Mathematical measurements and graphs have been employed in attempts to obtain a standard or central form among fossils, and a considerable measure of success has been obtained. *Trends* in evolution have been indicated by correlating changes in form with stratigraphical position, but the personal factor cannot be entirely eliminated. None-the-less such trends have in some instances been found concordant, and can be employed in stratigraphical determinations with some confidence.

Biology

Nor can the contribution of palaeontological research to biological studies be forgotten. The present animal and plant world is only a phase of organic nature as a whole. Biological theories must be checked by reference to discoveries among fossils, and in the realm of plant life, at any rate, the actual cellular structure of organisms, long since extinct, may sometimes be ascertained. The ultimate structure of biological remains may thus be studied just as the mineralogist may study the ultimate structure of crystals. In a few cases even cell contents, such as starch grains, have left traces, and it is claimed that nuclear material can still be distinguished, on occasion, from the ordinary cytoplasm of the cell. In the animal kingdom the cell wall is not so resistant as among plants, but, although the hard parts of skeletons are usually the only parts that are preserved, the soft tissues of the mammoth and woolly rhinoceros in the snow-drifts of Siberia are discovered from time to time. Flies and other insects in amber also retain their cellular structure. So the palaeontologist, too, can find scope for the study of the ultimate elements of the structure of fossil remains.

Geophysics

And both mineralogist and palaeontologist contribute to yet another geological problem of today—the present and past distribution of the continents of the earth. The idea that the continents are buoyed up on a plastic substratum dates back into last century; but it was only in 1915 that Wegener propounded his theory of continental drift. This has given a very satisfactory explanation of many of the difficulties confronting workers in all branches of geology, but disquieting difficulties are not wanting, as has already been indicated. But the geophysicist may discover, in the future, some factor that will suggest an array of forces sufficient to satisfy him that the lateral displacement of the continents is possible. The geologist can supply much evidence that is most easily explained on the hypothesis that such displacement has probably taken place.

Summary

In fine, then, the past hundred years has freed the geologist from the trammels of out-moded philosophies, and permitted him to make tremendous advances in every direction he cares to investigate. He may have made mistakes, but he is no longer confined and thwarted by intellectual taboos. He enters the year of the new Exhibition with a sense of satisfaction that he has contributed very largely to the present physical and intellectual comfort of mankind, and can look with confidence to a continuity of that contribution in the years ahead.

CHAPTER VIII

THE EARTH'S ATMOSPHERE

by P. A. Sheppard, b.sc.

From the seventeenth to the early nineteenth century almost every physicist was a meteorologist. The knowledge acquired in the laboratory was illustrated by such atmospheric phenomena as came within his cognisance, extended perhaps by climbing a mountain or, for the more adventurous spirit, by ascending into the atmosphere in a balloon. But little general structure emerged, nothing comparable for example with the corpus of astronomy. Halley, the Astronomer Royal and father of dynamical meteorology, had pointed the way to an alternative approach with his map in 1688 of the trade-winds and monsoons and an attempted explanation, but there were few who followed his lead to see the atmosphere and its workings 'broad and whole'. Those who did so—and Humboldt, the distinguished geographer, was almost the first at the turn of the eighteenth century—were regarded by most of their colleagues as rather unscientific map makers and imprecise observers. And so arose in the first half of the nineteenth century a regrettable alienation between the laboratory physicists on the one hand and the 'synoptic' meteorologists on the other, an alienation which has not everywhere disappeared today though, in an age of specialization, it is perhaps less discernible.

But the map makers were right to pursue their methods. They were the pioneers of a new technique demanded by meteorology, a technique different in kind from that of the experimental and theoretical physicist of the day and later. For they realized that meteorology is above all the study of weather, and weather is a large-scale, four-dimensional activity of a continuous fluid medium, only to be fully comprehended as such. That comprehension required the making and collecting of observations from wide areas and an effective method of presentation for their study.

H. W. Brandes, a mathematician of Breslau and Leipzig, was the first, in 1820, to discuss the weather as a whole over Europe day by day, and he later made a study of the life history of certain depressions by means of *synoptic* charts. Others quickly followed Brandes's lead, particularly Espy and Loomis in America and Dove in Germany, and their analyses led to an appreciation of the existence at or near sea-level of centres of high and low pressure and of winds circulating round them. Then too was born a theory of depressions (see below) which persisted until the close of the nineteenth century. The introduction of the electric telegraph in the 1840s provided the opportunity for a more ready dissemination of weather observations and Glaisher, of Greenwich Observatory, organized the first collection of daily reports by this means in Great Britain. They began to appear in the *Daily News* in 1849 and were published daily in mapped form at the Exhibition of 1851 and sold to the public at a penny each.

Meteorology in 1851—The Dawn of the Synoptic Approach

Let us now try to assess, rather shortly, the body of systematized knowledge and physical understanding of atmospheric phenomena against which these weather maps of 1851 could be viewed. The very mapping had, of course, taught men the existence of large-scale atmospheric systems, now termed depression, anticyclone, etc., and something of their behaviour. A rather well-defined relation was found between wind, pressure distribution in the horizontal, and weather. This relation was partly reflected in the rules of the weather glass, first introduced in simple form in the time of Hooke. Men also discovered that air did not move, directly at least, from regions of high pressure to regions of low in the horizontal, but that it kept low pressure on its left in the northern hemisphere and on its right in the southern—a rule a little later styled 'Buys Ballot's Law'. But the physical basis of these relations was hardly glimpsed.

In the lower latitudes of Halley's map the tropical cyclone had been identified by seamen naturalists and recognized as an intense circulation a good deal smaller in horizontal extent than the extra-tropical cyclone but with a similar (anti-clockwise in the northern hemisphere) rotation of the air. It was known to move rather slowly westwards, curving more and more rapidly into higher latitudes and occasionally emerging as a larger, less intense, eastward-moving depression in middle latitudes over the sea. Still smaller systems, such as the tornado and the cumulo-nimbus, fell within the limits of vision of a single observer.

Humboldt, following Kerwan, constructed maps of mean surface pressure and temperature over substantial parts of the globe early in the nineteenth century, while from his mountain observations and others' occasional manned-balloon ascents it became known that the temperature usually decreased upwards at a rate of about 1° C. per 500 ft. So was laid the basis for a theory of winds.

Pascal had observed that barometric pressure decreased with height, but it was Laplace (1805) who placed the phenomenon on a quantitative basis and provided meteorology with one of its basic laws—that the rate of decrease of atmospheric pressure with height is equal to the weight per unit volume of the air.

Cloud forms had been classified by Luke Howard, a London weather diarist, in 1803, and his classification remains essentially unchanged today. At the same time Dalton, the keeper of a weather diary for fifty-seven years, was making his famous experiments on the vapour pressure of water in the presence of air and on the condensation of water vapour by expansion to produce a cloud. He was thereby led to deduce that natural clouds were formed by the adiabatic cooling of humid air when ascending into regions of lower pressure—one of the most important inferences in the history of meteorology.

Early theories of weather systems—convection

As to the theories of weather processes in 1851, meteorology might almost be said to have been bedevilled by the dogma, given by Aristotle in his *Meteorologica*, that "everything that is warm tends naturally to rise". The significance of Laplace's equation, that pressure is the weight of the superincumbent atmosphere and so is imposed from above, and of the hydrodynamical equations of motion, that, gravity apart, the forces acting on the air are as much the result as the cause of its motion, was evidently not appreciated.

The first of the convectional theories to rear its head was that of

Halley for the trade winds and monsoons of low latitudes—warm air ascends in the hotter regions of the globe, causing a flow along the surface from the cooler regions on either side. Is this thesis still presented in some of the elementary texts as almost self-evident, even though it be overlooked that the equatorial parts of the earth are regions of minimum temperature difference—a fact known now for well over a hundred years? It is evident that meteorology in 1851 was not entirely happy about it, even after Hadley in 1735 and again Dalton independently had shown that the direction of such winds might be modified by the earth's rotation (right hand deflection in the northern hemisphere, left hand in the southern).

Espy's "Philosophy of Storms."

Convectional theories of tropical and extra-tropical cyclones or depressions were propounded between 1820 and 1840 by Dove, Espy and Loomis, who based their arguments upon Dalton's theory of cloud formation and the latent heat liberated when condensation occurred. Espy's views, expressed in his famous *Philosophy of Storms* (1841), are characteristic. He assumed that a rather extensive volume of relatively warm or moist air would rise through buoyancy, giving rise to cloud and rain, and would maintain its buoyancy and ascent while appreciable condensation continued by drawing upon its supply of latent heat. The ascent would draw in air at lower levels towards the centre of the system, which, so Espy argued, would be a centre of low pressure because the central column was composed of warm air, and air would be expelled from the system at high levels. The inflow would acquire an anticlockwise circulation (in the northern hemisphere) on account of the earth's rotation and the whole system would be carried along from west to east by the upper winds. This theory—surmise is a better term—persisted, though rather uneasily, throughout the century. But Espy and the others lacked a very important discipline only provided much later—an adequate series of observations, not merely at the surface but in the free atmosphere also.

The Part Played by Radiation

Let us come now to a matter always near to the heart of meteorology, the properties of solar and terrestrial radiation. It was appreciated that invisible radiation from the earth's surface as well as the solar beam helped to control the temperature of the earth's surface. In fact, Charles Wells in 1814 had given a theory of dew formation based on the nocturnal cooling of the earth's surface to below the dew point in consequence of the radiation emitted by the earth's surface to space. In doing so he gave an inkling of the reason for the nocturnal fall of air temperature above the ground.

But it was not appreciated before Tyndall's work in the 1850s that *gases* may absorb and emit radiation at room temperature and it was much later when water vapour and carbon dioxide were recognized as being active in this respect in the atmosphere. It is of interest to note here, in view of much later work, that W. Herschel in 1800 made the first attempt to examine the distribution of energy in the solar spectrum—though his results did not prove correct—and that he sought, inconclusively, for a relation between sunspots, as an index of solar activity, and the weather as evidenced by the price of corn.

Early Atmospheric Electrical and Auroral Studies

Another study, atmospheric electricity, aroused steady interest almost from the time that electrostatics itself began to take shape. Lemonnier found in 1752 that in fine weather the air near the ground is at a positive potential relative to the earth, and it was shown later in the century that the potential increased with height. Peltier then showed that the electric field could be regarded as arising from a negative charge on the surface of the earth. While Lemonnier investigated the electricity of fine weather, D'Alibard and Franklin made the first investigations on the electricity of disturbed weather and the lightning flash by means of pointed rods and kites respectively. But in 1851 the study of electricity, particularly in gases, had not progressed sufficiently for any explanation of the phenomena observed to be advanced.

The aurora had naturally attracted the attention of the scientifically minded from early days, and Hadley had discussed the relation between it and the earth's magnetic field in 1735. Dalton, who ranged so widely, made an admirable estimate of 153 miles for the height of an auroral display in 1793 but it is not clear whether by 1851 any inference had been made about the existence of an atmosphere at that height.

The Meteorological Society and world-wide observing stations

We may close this short survey of early meteorology by noting the creation of the Meteorological Society of London in 1823. It came into existence mainly to organize a world-network of observing stations so that meteorologists might better know the facts which required to be explained. As John Ruskin, one of its members, wrote in 1839:

"The Meteorological Society, therefore, has been formed, not for a city, nor for a kingdom, but for the world . . . so that it may be able to know, at any given instant, the state of the atmosphere at every point on its surface."

Seventy-one observing stations were founded over the globe, but the initial enthusiasm was evidently not maintained, for the Society became defunct in 1841. But a new Society sprang up in 1850 with rather less ambitious aims, and celebrated its centenary in 1950. Meanwhile Sir Edward Sabine, Secretary and later President of the Royal Society, was instrumental in establishing a number of magnetic and meteorological observatories in British possessions throughout the world.

International Meteorology and Weather Forecasting

The advantages to be gained by applying meteorological knowledge to marine navigation, as demonstrated by Maury of the U.S. Navy with his charts (1848) of ocean winds,[1] and the damage by hurricane to the Allied fleet in the Black Sea in 1854, led to the creation of a number of State meteorological services in the 1850s to collect and co-ordinate weather observations and provide advice for shipping. These services collaborated in the daily exchange of information and so were enabled to prepare large-area charts of current weather on which the distribution of surface pressure could be effectively displayed.

[1]Maury's charts led almost immediately to a very striking triumph of applied meteorology; the average sailing time from England to Australia and back was cut down from about 250 days to 190 days.

A quite short experience of these daily synoptic charts led a number of meteorologists to a perhaps too enthusiastic view of their immediate practical utility. Thus, on a quite empirical basis, weather forecasts and storm warnings were issued from 1860 by Buys Ballot in Holland, and, a little later, by Fitzroy in England. Fitzroy died in 1865, and in 1866 the Royal Society suppressed further issues in Great Britain and itself became responsible for the meteorological service. The storm warnings were restored in response to popular demand almost immediately, but forecasts were discontinued for twelve years.

Meanwhile the Royal Society, with Sabine and Galton as leading spirits, instituted the major enterprise of making autographic records of pressure, temperature, humidity, wind and rainfall, together with auxiliary eye observations, at seven observatories well-spaced over the British Isles. Facsimile reproductions of the records over a twelve-year period, with notes on the behaviour of each cyclone and anticyclone entering the area, were published, but they failed to elicit the general theory for which meteorology was waiting. In spite of this, forecasts, on slightly improved empirical methods of R. H. Scott and R. Abercromby, were recommenced in 1879.

By that time the system of high pressure had been brought into the picture of atmospheric circulation by Galton, who named it 'anticyclone'. He regarded it as a sort of inverted cyclone—a region of descending dense air which flowed radially away in the lower layers, acquiring clockwise rotation in the process. As with Espy's cyclone, it owed more to surmise than to theory, and gave no essential understanding of the genesis and history of the system, but the importance of vertical motion, in the one system as in the other too small for direct observation, is not to be questioned.

The events of the 1850s had created a spirit of internationalism among meteorologists. This spirit flowered in 1872 at Leipzig and in 1873 at Vienna, when at successive international conferences the representatives of a large number of national services, most of them providing forecasts for their publics, agreed on the forms of telegraphic messages and on instruments and methods for the standard kinds of observation, so that synoptic and climatological values might be made comparable. The International Meteorological Organization then formed has continued in being ever since.

Three-Dimensional Meteorology. The Systematic
Study of the Upper Air

Kite and balloon flights before 1851 were much too infrequent to provide either a climatology of the upper air, or, still less, an idea of the changes in vertical structure with change in weather. Flights continued to be made rather sporadically until about 1890 when a tremendous growth of activity occurred following the introduction of the box kite used with piano wire (capable of reaching 20,000 feet or so) and of the sounding balloon, both carrying self-recording instruments but no observer. By 1896 internationally co-ordinated ascents were being made, and at the turn of the century the epoch-making discovery of a region of almost uniform temperature, the *stratosphere*, beginning at about 10 km. height over Europe, was announced. The region below, of vigorous mixing and non-uniform temperature, was termed the *troposphere*.

This pioneering movement has a magic quality; as have the names associated with it—Rotch at Harvard; Teisserenc de Bort at the specially founded Observatory of Trappes; Assmann of Lindenberg (Berlin) who introduced the closed rubber balloon in place of the earlier, open-mouthed balloon of varnished paper; Hergesell, who went to sea with the new techniques and organized the international upper air ascents; and W. H. Dines, the pioneer in Great Britain, who invented an ingeniously simple and light meteorograph, in which pressure, the meteorological variable, was first used instead of time as the basic co-ordinate of the record.

The reality of the stratosphere was at first questioned but there was much more reason for surprise at two other features which came to light, the often very sharp discontinuity in *lapse-rate* (or rate of fall of temperature) between the troposphere and stratosphere—a transition (termed the *tropopause*) yet to be explained— and the fall of temperature in the stratosphere with decreasing latitude. With that decrease goes an increasing depth of troposphere. Moreover, it was shown, mainly by Dines, that the height of the tropopause increased and its temperature decreased with rise of pressure at the surface and with rise of the mean temperature of the troposphere. So was the behaviour of the atmosphere at one level correlated for the first time with its behaviour at other levels.

In the 1930s upper-air sounding took a big step forward with the introduction of the *radio-sonde*, a cheap but reliable device for signalling to the ground the pressure, temperature, humidity and, latterly, the wind at the levels through which it passes. With this tremendously powerful tool and the will to use it—the Second World War provided an effective spur—there has been a rapid accumulation of knowledge of the ways of the atmosphere up to about 20 km., that is over more than 90 per cent of the mass of the atmosphere in a vertical column. The dreams of some of the pioneers are in process of realization, for we are now able to formulate some of the main problems of atmospheric behaviour. And the immediate reason for this expansion in knowledge has been, in the main, to provide the forecaster with additional current data.

Developments in Synoptic-Chart Analysis. The Structure of Weather Systems

Experience with synoptic charts led Abercromby in 1885 to put forward the structure shown in Fig. 1 as representative of a

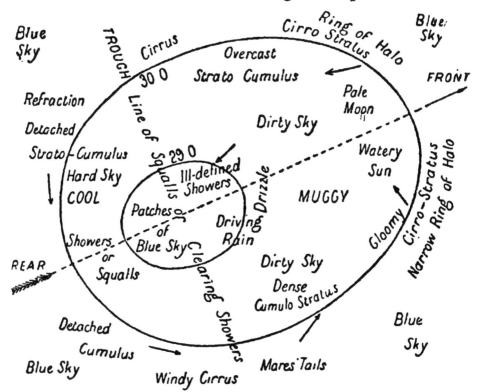

Fig. 1. Prognostics and weather of a cyclonic depression according to Abercromby (1885).

depression. The most notable addition to the pre-existing picture was the *trough-line* separating, on the south side, an area of steady rain to the east from an area of showers on the west; warm, moist air from cool air; and a relatively weak horizontal pressure-gradient from a stronger (as indicated by the separation of the isobars). It will be noted that there is no evidence here for the existence of a warm core at the surface, as demanded by early theories. The first clear effort, however, in the elaboration of the processes in a depression was made by Shaw and Lempfert in 1906 in their classical paper, *The Life History of Surface Air Currents*. We may only note here that they found characteristic differences in the patterns of cloud and rain and of the trajectories of the air entering and leaving slow and fast moving disturbances, but that there was in general a confluence of currents which Shaw represented by Fig. 2. He supposed the rainfall to result from convergence within the warm damp current, the ascent of this air over a colder current to the north and its undercutting by a colder current from the west.

This description stood until J. Bjerknes put forward in 1918, and with Solberg in 1922, the so-called *polar-front* or *wave theory* of cyclonic depressions. It builds on Shaw in showing the depression as a region of interaction of warm and cold air masses, but fastens attention on the *life history of the system itself*. Bjerknes's picture, a piece of inspired descriptive meteorology based on the synoptic chart rather than a theory, shows a depression forming at a northward bulge in the gently sloping boundary surface or *front* separating air masses of widely different origin and properties. Under certain conditions, not then clearly understood, the tongue of warm air (or *warm sector*) constituting the bulge grows in size, slides up over the underlying cold air ahead of the tongue and is undercut by the cold air behind it. These stages are shown in Fig. 3 (*a*) to (*c*). During them the system moves slowly in the direction of the flow in the warm sector and acquires a set of closed isobars centred on the tip of the warm tongue. There is a broad belt of rain ahead of the warm sector and a narrower zone of showery rain at its rear. The depression then begins to *occlude*, that is the (*cold*) front on the rear side of the warm sector overtakes the (*warm*) front on the forward side, lifting the warm air completely from the ground, as in Fig. 3 (*d*) to (*f*).

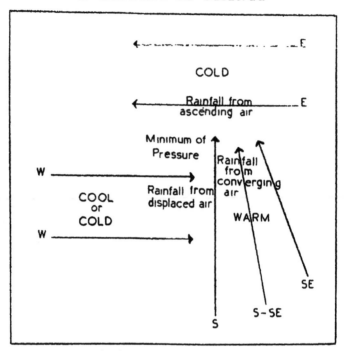

Fig. 2. The horizontal distribution of wind, temperature and rain with reference to the centre of a cyclonic depression, according to Shaw (1911).

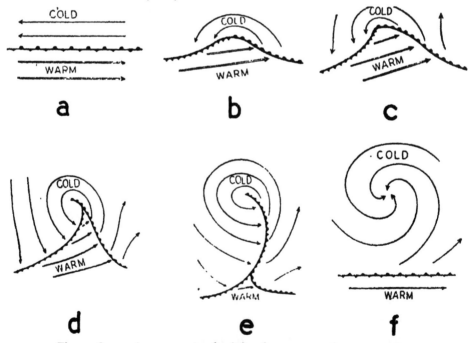

Fig. 3. Successive stages in the life of a wave cyclone according to Bjerknes and Solberg (1922).

Occlusion is the most active stage in the life cycle; in it the system acquires a great increase of kinetic energy, it moves most rapidly and the precipitation is heaviest.

The ultimate state, though there are variants from this, is of a more or less thermally symmetrical vortex of cold air, which slowly dissipates through friction. Vertical sections through the warm sector and to the north of its tip, showing cloud systems and precipitation in the unoccluded stage, are shown in Fig. 4.

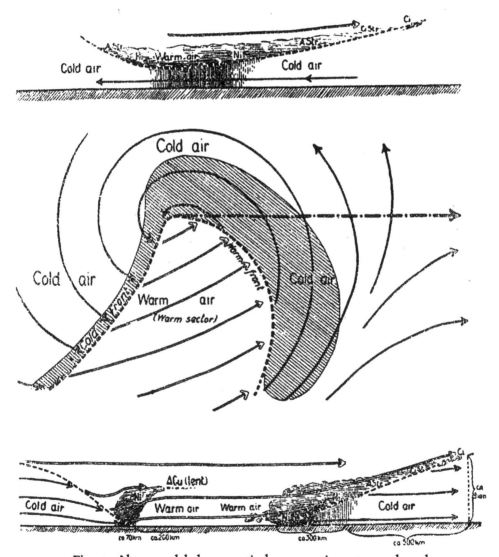

Fig. 4. Above and below: vertical cross-sections, to north and south respectively, of the centre of an idealized wave cyclone shown (centre) in horizontal section, after Bjerknes and Solberg.

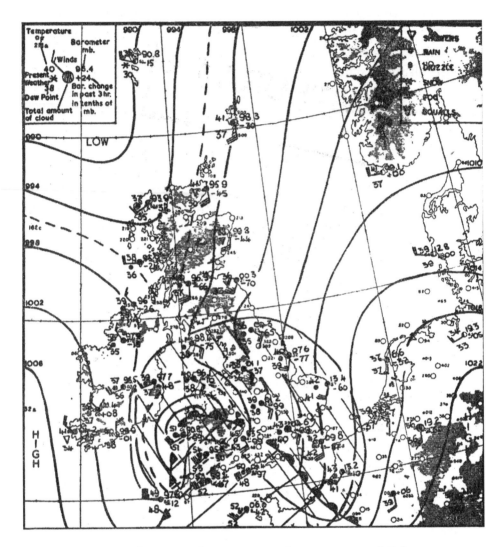

Fig. 5. Wave cyclone centred over S.W. Wales at 6 a.m. on 9th February, 1949. The associated areas of rain are shown shaded. The system has formed at the extremity of a cold trough extending S.E. from Iceland to the British Isles.

An actual synoptic chart showing a particular stage in the process is given in Fig. 5, and, over a larger area, in Fig. 6. A vertical cross-section through the atmosphere along a line shown in Fig. 6 is reproduced in Fig. 7; this is notable for a very concentrated pattern of high wind which has come to be called a *jet stream*.

Points of similarity between the various descriptions of the depression, from Abercromby to Bjerknes, will be evident to the reader. For example, Abercromby's trough line is the line of separation of warm southerly from cool westerly air in Shaw's picture, and is the cold front or occluded front of Bjerknes's system. If in Shaw's diagram the arrows indicating easterly wind are joined up to the arrows indicating westerly wind, the similarity with Fig. 4

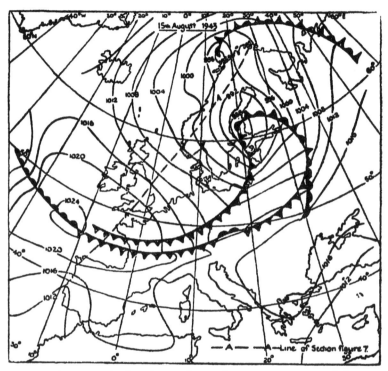

Fig. 6. The synoptic situation at the surface over Europe and the East Atlantic on 15th August, 1943, showing associated wave disturbances in various stages of growth and decay. Thin lines are isobars; heavy serrated lines are fronts—warm fronts rounded serration, cold fronts pointed, occluded fronts rounded pointed; for the dashed line AA see Fig. 7. (After Durst and Davis).

is great; but Shaw was not ready to take this step as he found the air masses to north and west of the system to be often of different origin.

Since Bjerknes's original papers, the structure of a depression has been related to upper air conditions, and it has been found that the axis of the system, taken as the locus of the centre of low pressure at successive levels, slopes backwards, i.e. westwards, and gives place, at a higher level the older the depression, to an open wave-shaped system of isobars, whose wave-length is normally much

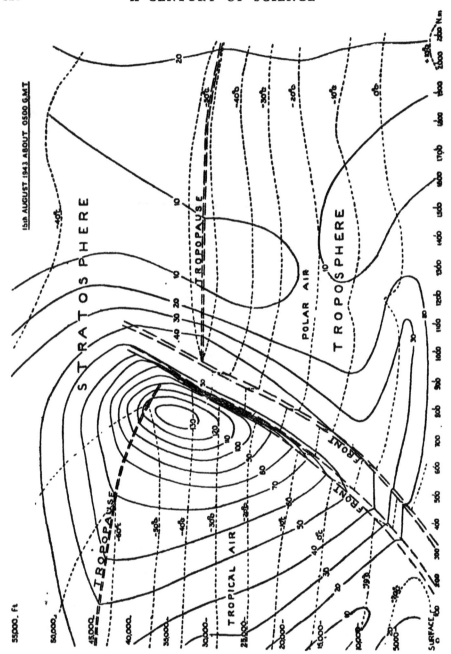

Fig. 7. A vertical cross-section through the atmosphere along the line AA of Fig. 6. Distance north-west along AA at bottom. Isopleths of wind speed in knots, normal to the section, are shown by full lines; those of temperature (° C.) by broken lines. Fronts and tropopauses are shown by double broken lines. The section shows a well-developed pattern of wind with a strong maximum beneath the tropopause in the warm air over England. The concentrated region of high wind, running N.W.—S.E. in the case illustrated, is termed a jet stream. (After Durst and Davis.)

greater (*c.* 3,000 km.) than the 'wave-length' of the system below (*c.* 1,000 km.).

The physical processes responsible for the life-history of a depression are matters which have naturally received much attention since Bjerknes's original papers. This work has been mainly along the lines of determining the conditions for dynamical instability of a mass of air subject to an initial disturbance. Factors such as critical wave-length and growth rate have been determined, but the values obtained depend naturally upon the properties of the model assumed to be representative of the atmosphere, and here there continues to be ground for debate. We shall return to this problem, and shall find the depression to be related with other characteristic weather systems.

The Present Picture

We must now attempt to put our study of weather systems into some relation with other properties of the atmosphere as now known and understood, and contrast it with the position in 1851.

Radiation and Temperature

There is little progressive change, year by year, in the temperature distribution over the globe; that is, the atmosphere is in a state of very nearly thermodynamic balance.

The temperature distribution at the earth's surface was known in broad outline in the middle of the nineteenth century and has lately been filled in for the upper air, a recent representation elaborating some of the features referred to earlier being given in Fig. 8. This distribution is the outcome of radiation and atmospheric motion. By radiation alone the lower latitudes would continually heat up and the higher latitudes cool down, and the atmospheric circulation is brought into being by the gradient of temperature so as to transport heat polewards and maintain the gradient in a quasi-steady state. The details of this process form the heart of meteorology. Atmospheric radiation probably cools the troposphere at all levels but more aloft than below, and the motions of the atmosphere —small-scale systems (commonly called *turbulence*), convection, and larger-scale patterns—are again responsible for maintaining a balanced state, this time in the vertical. But we are still rather ignorant

I

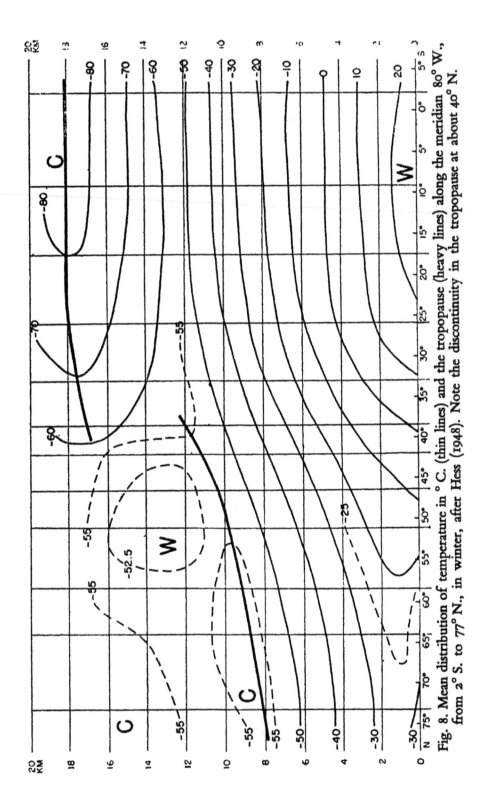

Fig. 8. Mean distribution of temperature in °C. (thin lines) and the tropopause (heavy lines) along the meridian 80° W., from 2° S. to 77° N., in winter, after Hess (1948). Note the discontinuity in the tropopause at about 40° N.

of radiation processes in the atmosphere, of absorption from the solar beam at high levels, and of atmospheric radiation at all levels, and the meteorologist is waiting for the physicist to provide him with definitive data on the radiative properties of gases *under the conditions of pressure and temperature in which they exist in the atmosphere.*

Cloud Physics

We turn now to the physics of cloud formation. A century ago it was realized that clouds were due to the more or less adiabatic ascent of moist air, and since that time their patterns have been related to the nature and scale of the process. But only in the last twenty years or so has it been realized that the co-existence of ice and supercooled water in the upper reaches of clouds is generally necessary in order that a cloud may precipitate its moisture. Aitken pioneered the micro-physics of liquid condensation on nuclei in the 1880s, and Findeisen of sublimation of ice in the 1930s, following Bergeron's thesis that a cloud becomes colloidally unstable when ice particles are formed. Such particles grow at the expense of supercooled water droplets in their neighbourhood because of the lower saturation vapour pressure over ice, but the details of ice formation and growth are not yet understood and are the subject of active research at the present time. So we have still to determine the immediate causes of rain, hail and snow.

Atmospheric Circulations

The aim of dynamical meteorology is naturally to obtain such an understanding of atmospheric processes that, given the state of the atmosphere in sufficient detail at any one time, its behaviour thereafter (in regard to wind, cloud, rain, temperature, etc.), may be predictable with reasonable accuracy for a considerable time ahead. We must record a glorious failure to do this very thing by L. F. Richardson in his classical *Weather Prediction by Numerical Process* of 1922. Richardson failed for various reasons but he pointed the way ahead. Whether, however, the aim, as commonly imagined, is even realizable is not yet certain. For with a medium whose interesting behaviour arises so much from instability in one form or another, it may be necessary to pose the problem in terms of probability, the question to be asked depending on the length of time for which an answer is required.

The problems of dynamical meteorology are problems in scales of motion. Thus we already know that the atmosphere possesses a large number of natural modes of motion, varying in scale from the micro-turbulence of the frictional boundary layer through the motions of convection and the tropical cyclone to the extra-tropical depression and anticyclone, 'long waves in the westerlies', and so forth. What modes it may possess on a quite large time-scale we do not yet know for certain, but climatic variations, with 'periods' up to the aeons of the ice ages, are a reality; what we do not yet know is whether such variations are entirely 'forced' by external agencies such as solar variation, or whether they too are natural modes of oscillation, of earth (land, water, ice) and atmosphere together.

Some atmospheric modes, for example, fine-weather cumulus and the tides, are stable; energy-producing and energy-dissipating forces being in balance. They excite less interest than the unstable modes, which result from an initial perturbation, grow and pass through a characteristic life history, ultimately being degraded by smaller-scale systems and friction. Rossby and his co-workers have demonstrated some features of the life-history of long waves in the westerlies, showing how they may give rise to cut-off pools of cold air in relatively low latitudes, while the Bjerknes cyclone is an illustration on a smaller scale and the cumulo-nimbus (thunderstorm) on a yet smaller. Eady has shown in the first two cases that the instability and life-history derive from the meridional temperature gradient and the vertical stability of the atmosphere, both systems being modes of atmospheric overturning—not in a vertical plane—which bring about a re-distribution of heat in the required sense (cf. *Radiation and Temperature* above).

Thus a measure of co-ordination is appearing among the systems which are to be seen on the synoptic charts of extra-tropical latitudes. There is moreover a vague understanding of the circulations from a thermodynamic standpoint, of heat absorbed into the system at lower latitudes and lower levels, and discharged aloft, by radiation. But the motions in low latitudes remain to be co-ordinated with those in moderate and high latitudes and the problem of secular variation in the circulations, that is of climatic change, is almost untouched.

The Exploration of the High Atmosphere

Since the First World War several methods of atmospheric exploration have been developed which provide information on levels up to about 1,000 km., much above the upper limit (about 30 km.) of the sounding balloon. These methods comprise: the brightness and movement of meteor trails, the propagation of sound over great distances by refraction at heights of 30 to 50 km., the measurement of electron density and stratification by radio-wave reflection from the ionosphere, the vertical distribution of ozone, the spectral analysis of aurorae and the light of the night sky, atmospheric tides and terrestrial magnetic variations, noctilucent clouds, and the variation of the intensity of twilight with time after sunset.

We cannot here go into the details of these techniques nor into the results which each technique individually provides. We can, however, take note of the general picture for not-too-high latitudes which they are mainly consistent in providing, and which has been supported by the results of a few direct soundings by highly powered rockets. The temperature begins to rise appreciably from about 30 km., reaching a maximum greater than ground temperature at about 55 km. There is then a fall of perhaps 140° C. to a second minimum at about 80 km., above which the temperature rises through the successive layers of the ionosphere to very high values, perhaps 1000° C., at a few hundred km. Substantial diurnal variations may occur in the upper reaches of the first warm layer and in the ionosphere.

The composition of the atmosphere probably remains substantially unchanged, so far as the 'permanent' gases are concerned, up to the level of the second temperature minimum, where molecular oxygen begins to give place to atomic oxygen, the transformation being complete at perhaps 120 km. Of the non-permanent gases, ozone increases in amount from the lower stratosphere up to about 23 km., above which its density diminishes rapidly. The total amount of ozone increases with latitude, and from autumn to spring, and varies by up to 30 per cent from day to day in sympathy with air mass changes in the troposphere, probably from dynamical causes. The total amount and distribution of water vapour in the stratosphere are not yet known, but observations show that the relative humidity of the lower stratosphere over England is often very low— the absolute humidity is necessarily very low. From such information one may attempt to account for the observed temperature distribution aloft, though this remains mainly as an exercise for the future. At present it seems probable that the high temperatures of the ionosphere are due mainly to the absorption of the solar ultraviolet by oxygen, and those of the first warm layer to absorption by ozone, dynamical processes perhaps modifying the distribution. For winds, up to 100 metres a second or more, are known to occur in various layers up to 300 km.

Modern Atmospheric Electricity

We conclude with a short appraisal of the changes in view concerning atmospheric electricity. It became apparent in the second half of the nineteenth century that air is not a perfect insulator and that its conductivity varies in different situations. It was in fact from the study, at the turn of the century, of the natural conductivity of air by C. T. R. Wilson and others that the subject of cosmic rays was born. The ion production and conductivity were found to increase with height quite rapidly, except near the surface, while the fine-weather electric field diminished with height. The decreasing field implied that the atmosphere possessed a net positive space charge, and the electrical structure to be explained was that of a sort of condenser, of which one plate was the earth's surface holding a bound negative charge. This charge was equal and opposite to the space charge contained in the atmosphere above, which constituted the other, rather diffuse sort of plate. But it was a leaking condenser in which the ionic current, the product of the field strength and the conductivity, was sufficient to cause its discharge in less than an hour.

The problem to be answered was then: How is the condenser maintained in a more or less steadily charged condition? Several theories, now of historical interest only, had been propounded when Wilson suggested, about 1920, that the source of maintenance was to be sought in regions of disturbed weather, particularly of thunderstorms. He has proved correct; the thundercloud has been found to be an electrical generator within which charge is separated, positive towards the top and mainly negative towards the base. The field above and below the cloud is then the reverse of the fine-weather field and is much greater in magnitude, so that it is able to inject a large amount of positive charge into the upper, highly conducting, atmospheric levels and remove it, by point discharge and lightning, from the earth's surface. Wormell showed that the balance-sheet of charge transference over areas of alternately fine and disturbed weather could in fact be balanced, while Whipple showed that the diurnal variation of the potential of the upper conducting layers was in close sympathy with the diurnal variation of integrated thunderstorm activity over the whole globe.

The major problem which now awaits solution is the mechanism by which electricity is separated in the thundercloud. Simpson and his collaborators showed that the separation occurs mainly in the glaciated regions of the cloud, but no adequate theory has yet been offered. We return here to problems in cloud physics and to the importance of ice which we have already noted in discussing the growth of precipitation elements. The two problems are closely related and they provide a happy example of the interplay of widely different branches of physics, which is one of the attractive features of the study of the atmosphere.

CHAPTER IX

THE CONSTITUTION AND EVOLUTION OF STARS

by W. H. McCrea, M.A., Ph.D.

In this chapter the 'constitution' of a star is taken to mean principally that of its main interior, and its 'evolution' to refer principally to its career as an independent body. With this interpretation, our subject is the central one of cosmical physics. For a star is the least body of matter that may be treated as uninfluenced by other matter over a long interval of time; only after understanding the constitution of a star can we hope to understand the behaviour of other astronomical systems.

The subject is particularly appropriate for the present review. A hundred years ago it was just beginning to be possible to formulate the problems. And now it seems allowable to claim that what have since come to rank as the most fundamental of these problems have been solved during the concluding years of the century. This is perhaps the crowning achievement of the century in physical science. For it has explained the properties of a star in terms of those of the fundamental particles of laboratory physics, and in doing so it has employed every fundamental advance in pure physics and has depended upon observations made with the aid of all the resources of physical technology.

Just because the subject embraces so much, it is a difficult one to treat concisely. Also its special concepts may be less familiar to the reader than those of other chapters. It seems best, therefore, to devote most attention to a description of the present state of knowledge. This is given immediately after a brief description of the state in 1851. The subsequent summary of the intervening development should then be more readily intelligible, though it has to be left to the reader to relate it to other simultaneous developments recorded in earlier chapters.

State of Knowledge in 1851

In 1851 the masses of a few stars had recently been determined and their luminosities estimated in terms of the Sun's luminosity. Also, nearly thirty years previously, J. Fraunhofer had examined the spectra of several stars and noted differences and similarities between them and the Sun's spectrum. From such evidence, and not merely from general inference, the stars had been shown to be bodies of a sort generally similar to the Sun. But it was also appreciated that their range of intrinsic luminosities must be large.

Very little was known, however, about the physics of the Sun. Some attempts had been made to measure its thermal power. Fraunhofer had mapped its line-spectrum in 1814, but by 1851 the origin of the Fraunhofer lines had not been established, though a few hints had been obtained. Physical observations of the Sun had been largely confined to sunspots and associated phenomena; the solar prominences and corona had been the subject of qualitative observations during eclipses. (It was only as a result of observations made at the eclipse of 1851 itself that astronomers became generally agreed that the latter appearances are genuinely part of the Sun.) But these observations concerned phenomena that still present intractable problems and that in any case have scarcely helped towards understanding fundamental solar constitution.

The view of the Sun's constitution was still substantially that enunciated by William Herschel in 1795. The Sun's radiation was supposed to come from a self-luminous envelope ('photosphere') overlying a cooler ('planetary') atmosphere, while the main body of the Sun was supposed to have a cool, solid, and probably habitable surface.

In 1848, J. R. Mayer had concluded that the Sun's radiation does not come from mere cooling nor from ordinary chemical combustion and had advanced the hypothesis that the heat needed for the radiation is produced by the bombardment of the Sun by meteoritic matter from space. He gave calculations showing the advantages and possible difficulties in this suggestion. Though this work attracted little notice it was highly significant as the first formulation of an astrophysical problem in terms of the new ideas of the conservation and convertibility of energy. The hypothesis was advanced independently by J. J. Waterston (1811–83) in 1853 when it received much more attention.

Although so little was known of the physics of extra-terrestrial matter, the first half of the nineteenth century had witnessed the greatest advances ever made in physics itself in the discoveries of the atomic structure of matter, the conservation of energy, and the laws of electromagnetism. These provided the foundation for the tremendous simultaneous progress in physics and astrophysics which quickly followed and which have yielded the results we shall now attempt to describe.

Physics and Chemistry of the Sun

In order to assess the present state of knowledge of our subject, we must first ask, In what circumstances can we say that we have understood the constitution of a star? We begin with the case of the Sun: let M_0, R_0, L_0 be its observed mass, radius, and luminosity.

Consider then a quantity of matter of mass M_0 (to be called *the* matter) left to itself remote from all other matter. Suppose we deduce what kind of a body the matter will form and use for our deductions nothing but known laws of terrestrial physics. Further, suppose we find that the deduced body is also a sphere of radius R_0 and that it also emits radiation at the rate L_0. Then we can almost certainly say that the constitution of the Sun is the same as that of this body—that this body is in fact the Sun. We should have demonstrated that physical laws require the Sun to be what it is: such is all that we can mean by a physical understanding of the Sun.

We shall now see how far the deduction can be carried at the present time.

The matter under consideration must be taken as composed of chemical elements in some definite proportions. Part of the object is to see how the results depend upon this composition, and we shall return to this point.

To get a definite picture, let us suppose the matter to be initially in a diffuse state and then to begin to draw itself together by its own gravitation. Such contraction must lead to collisions between different parts of the matter whereby the gravitational potential energy lost in the contraction would be converted into heat-energy in the matter. In due course some of this heat would be lost by radiation into surrounding space. This suggests that the contraction must proceed indefinitely.

Suppose then the contraction to have proceeded until the volume of the material is such that the average density is one or two grammes per cubic centimetre. The pressure by which the matter must necessarily tend to oppose its own contraction can be estimated and its average value for this mass and volume is found to be of the order of 1,000 million atmospheres. Now matter of this density can exert such a pressure only if it is very hot indeed. Its atoms must then be stripped of most of their electrons and thereby greatly reduced in effective size. The matter can therefore behave like a gas even if its density is large. Atomic theory puts all this into well-known formulae which show that the matter must be in fact a 'perfect' gas at a temperature of some millions of degrees.

One consequence of the gaseous state of the matter is that by this stage the whole mass must have formed itself into a spherically symmetrical body (unless it were initially endowed with considerable rotation, which we suppose not to be the case). This body will *tend* to go on contracting, with continually rising internal temperature. Of the gravitational energy liberated throughout the body by the contraction, part will go into this temperature-increase, and part will be radiated away from the boundary. The latter part must first be transported from the interior. The physics of matter in this state shows that the means of transport is predominantly a net outward flux of radiation, rather than convection or conduction. Also it shows that the opacity of the material to radiation is due partly to the photo-electric effect of the electrons still bound in the depleted atoms and partly to scattering by the free electrons; it provides formulae for this opacity.

Owing to the outward energy-flux, the temperature is greatest at the centre of the mass. When, in the course of the contraction, this temperature reaches about 20 million degrees, new considerations arise. The thermal agitation of a highly ionized gas causes the atomic nuclei perpetually to suffer collisions with each other. At temperatures lower than 20 million degrees, the nuclei are virtually unscathed by these collisions. But at about this particular temperature an appreciable proportion of the collisions become so energetic that they cause the nuclei taking part actually to react with each other.

We now get transmutation of the chemical elements by such *thermonuclear reactions*.

The Carbon-Nitrogen Cycle

These reactions release sub-atomic energy. They can be described as sub-atomic 'combustion'. The ones releasing by far the most energy are those consuming hydrogen nuclei (protons) and it is those for which the temperature mentioned is particularly significant. Nuclear physics (see Chapter I) shows fairly conclusively that the end-product, or 'ash', of such proton-combustion is always helium nuclei. These can almost certainly be synthesized directly by successive combinations of protons. But a much more efficient process under the conditions we contemplate is one in which carbon and nitrogen nuclei act as catalysts. The rate of energy-release by this 'carbon-nitrogen cycle', as the process is called, is given by a formula which shows that, besides being proportional to the quantities of protons and catalysts available, it is also nearly proportional to the temperature raised to about the twentieth power. This extreme sensitivity to temperature is of the utmost significance, since it means that equal amounts of energy can be released by vastly different quantities of the reagents reacting at only slightly different temperatures.

Therefore, when the central temperature of our contracting body reaches about 20 million degrees, sub-atomic energy begins to be released in this way, provided the necessary ingredients are present. The contraction must be arrested when the rate of energy-release balances the rate of energy-loss by radiation from the boundary, because further contraction would raise the internal temperature and so would increase the rate of energy-release beyond the rate of energy-expulsion; the heat then accumulating would cause the body to re-expand. So the state attained is stable. Since the ability of the body to expel energy outwards will not be greatly affected by the consumption of protons, the body will remain in approximately the same state until the supply of protons for the thermonuclear reaction has been exhausted, the diminishing supply being compensated by an only slightly rising internal temperature.

Mathematical treatment bears out these inferences. In particular, it confirms that a stable configuration is achieved, but shows further that this results in the innermost 15 per cent, or thereabouts, of the mass being in convective equilibrium while the remainder is in radiative equilibrium.

The body is now a 'star' in that it is in a stable state and has a practically constant luminosity maintained from its own resources. For a considerable range of possible chemical compositions of the matter, the radius and luminosity calculated for this state are approximately those of the actual Sun. Also, with any of these compositions, the body would maintain about the same luminosity for many times the estimated past life of the Sun.

These results allow us to claim that we now know in main essentials the physical constitution of the Sun.

Some astrophysicists would go further and assert that it is possible to infer the main quantitative features of the chemical composition of the Sun from an exact agreement between theory and observation. This is questionable because of the insensitivity of the calculated radius and luminosity to the assumed composition and because of certain approximations in the computations. However, all calculations show that hydrogen must be present much in excess of all other elements.

Stellar Constitution

The foregoing discussion would apply to any other sufficiently large mass M of material isolated in space. Thus, provided it contains the requisite protons and catalysts, it will form a star on the general model of the Sun.

The luminosity L and radius R can be calculated for any value of M and any assumed chemical composition. The calculation shows that the relative importance of the different mechanisms of opacity depends upon the value of M and also that, for values of M several times the solar value M_o, radiation pressure has a significant effect. The incidence of these features depends upon the chemical composition.

For all such model stars of any prescribed composition there is therefore a precise relation between L and M—a theoretical *mass-luminosity law*. For different compositions, the relations are different, though qualitatively similar.

Correspondingly, there is a precise relation between R and M.

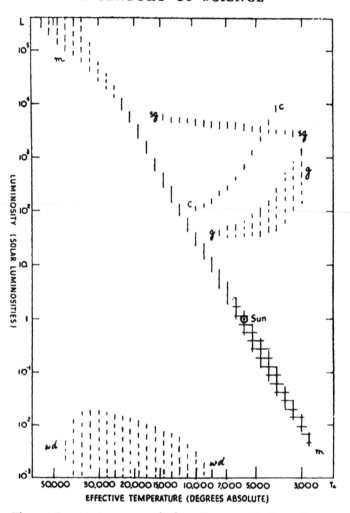

Fig. 9. *Luminosity—spectral class diagram.* Shading shows schematically the portions of the diagram containing the majority of points representing actual stars. *m,* main sequence; *wd,* white dwarfs; *g,* giants; *sg,* supergiants; *c,* cepheid variables. The doubly shaded region contains the most abundant main sequence stars. The effective temperature indicates the spectral class.

Taking the two relations together, there is a precise relation between L and R. But the radius and luminosity of a star together determine its 'effective temperature' T_e, i.e. the temperature of a perfect radiator which would emit the same total radiation per unit surface area as the star itself actually emits. The effective temperature predominantly determines the spectrum of the emergent radiation, i.e. the 'spectral class' S of the star. Therefore the relation between

L and R implies a relation between L and S—a theoretical *luminosity-class* relation. Once again, the relation depends somewhat on chemical composition.

Main Sequence Stars

If the luminosity is plotted against the spectral class for all stars for which these have been observationally determined, the resulting diagram reveals a remarkable phenomenon. Allowing for accessibility to observation, about 90 per cent of the stars in the region of

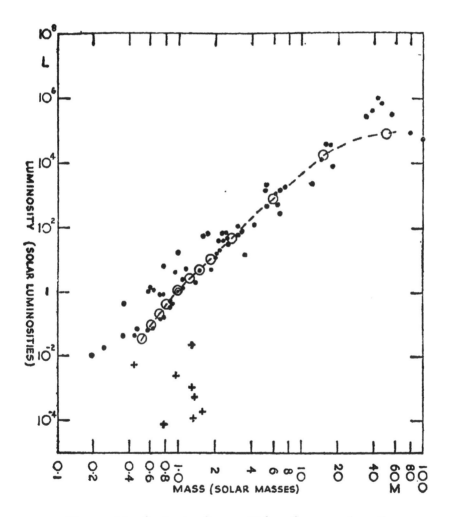

Fig. 10. *Mass-luminosity diagram.* Values for a number of actual stars are shown by dots for main-sequence and giant stars and by crosses for white dwarfs. Average values for main-sequence stars are shown by open circles.

space concerned are represented by points crowded into a single belt stretched obliquely across the diagram. Hence the great majority of stars belong to a single family: it is called the *main sequence*. A curve drawn in the diagram to lie centrally along the sequence defines an *empirical luminosity-class relation* to which all members conform to within a certain degree of scatter. The Sun is a fairly average member of the sequence. (See Fig. 9.)

Again, if luminosity is plotted against the mass for all main sequence stars for which observational values are available, this new diagram reveals another remarkable phenomenon: the stars are again represented by points scattered near to a single curve. This defines an *empirical mass-luminosity relation*. (See Fig. 10.)

These two phenomena are remarkable when discovered as purely empirical regularities amongst such a large proportion of the stars. But we now see that their existence is demanded by the theory. Moreover, it is fair to state that the agreement between the theoretical and empirical relations is satisfactory. The 'scatter' in the observational results can be explained by differences of chemical composition. But it can also be inferred that probably all except the brightest of the stars concerned contain 90 per cent or more by mass of hydrogen.

This is not to imply that the theory is already worked out in all its details. On the other hand, there is corroborative evidence of its general correctness beyond what can be given here. Most astrophysicists would agree that, although developments in atomic physics may produce modifications and the calculations may require elaboration, the principal properties of the main sequence stars are now understood.

White Dwarfs

The state in which we left our theoretical 'star' evidently placed it in the main sequence. Its luminosity was being produced by the transmutation of protons. In course of time, all the protons in the interior will be consumed. The form of the mass-luminosity relation shows that this must happen much sooner for the more massive than for the less massive stars. The contraction must then be resumed. The material will remain highly ionized; but the quantum theory shows that when the density becomes great enough, the material will no longer behave as a 'perfect' gas. Instead the free electrons will form what is known as a 'degenerate' gas.

The characteristic property of such a gas is that its pressure depends only on its density. Therefore, when the material of a 'star' of given mass has become wholly degenerate, the star will have attained a configuration beyond which no further contraction is possible. Since it is the degeneracy of the *electron* gas that is mainly significant, the atomic nuclei will for a time release thermal energy, and also sub-atomic energy from any transmutations still proceeding. When all such energy-supplies have been exhausted, the 'star' will remain as a cold (non-luminous) body.

Hence we infer the possible existence of stars of great density, having a determinate configuration for any particular mass, but of indeterminate small luminosity.

The well-known *white dwarfs* are stars having just this character. In the two diagrams above, they are represented by points lying well below the main sequence. On account of their intrinsic faintness, such stars have to be comparatively near us to be observed at all. Though only some twenty are actually known, it is estimated that about 10 per cent of all stars (with which this chapter deals) are white dwarfs.

Calculations show the theory to give a satisfactory account of the constitution of white dwarfs, in particular of their enormous mean densities, which are estimated to exceed 10 tons per cubic inch in some cases. Thus some 10 per cent of stellar matter is shown to be in a state that can be described only with the aid of quantum theory: attempts to account for white dwarfs before the discovery of quantum 'degeneracy' had led to a paradox.

The process responsible for the luminosity of white dwarfs probably offers no fundamental difficulty but has not yet been definitely identified.

(In the present account it is permissible to ignore certain elaborations which result from the consideration of 'relativistic' degeneracy and which are believed to be required under certain conditions.)

Giant stars

A majority of naked-eye stars are actually not main-sequence stars but are *giant* or *super-giant* stars. However, this is only because the giants are so much more luminous than average main-sequence stars. In a given region of space, the latter outnumber the giants by probably more than 100 to 1.

The giants are found to have masses corresponding to their luminosities in conformity with the empirical mass–luminosity relation, so that in this respect they do not differ from main-sequence stars. The difference is in their much larger radii, a typical giant having a mean density several thousand times less than that of the Sun. In fact, a normal main sequence star of large mass surrounded by and illuminating an extensive gaseous envelope would reproduce the characteristics of a giant.

Why a small percentage of stars should depart in this manner from what has come to be regarded as the normal behaviour represented by the main sequence is not fully understood. At present, the most promising suggestion relates to a feature not yet mentioned in our discussion. We have been treating the chemical composition as something characteristic of a star as a whole. But the tendency of thermonuclear reactions is to produce a variation of chemical composition between the central and outer zones. Now, it has been shown that such a variation would produce a larger theoretical radius than that for a star of uniform composition. Whether this effect will explain the huge distension of giant stars can be decided only by calculations still being performed.

Variable Stars

The stars most obviously presenting problems essentially different from those of the categories already described are the stars of observably variable luminosity.

Save for the relatively rare *super-novae*, the most remarkable of these are the ('ordinary' or 'galactic') *novae* or 'new stars'. The nova outburst is a sudden temporary increase of luminosity to perhaps 10,000 or 100,000 times its previous value. This is the most extreme form of 'irregular variability'. But some novae have been observed subsequently to behave like other and less extreme types of *irregular variables*. These in turn appear possibly to have something in common with certain stars that are observed to be continuously ejecting matter, though not detectably varying in luminosity.

It is coming to be believed that many of these departures of a star from a steady state may have a common origin. Though not yet fully elaborated, the most plausible theory is that they are associated with the collapse of a star from the main-sequence state towards the white-dwarf state, the phenomena occurring in any particular case being conditioned by the amount of rotation of the star concerned.

Stars varying rhythmically with a definite period present an altogether distinct set of phenomena. The most interesting are the *short-period variables*, commonly referred to as *cepheids*. Their most famous characteristic is an empirical relation between the period and the median luminosity. Since cepheids are stars of great intrinsic luminosity and so can be seen at great distances, this property has made them indispensable for establishing a scale of distances in the universe. From the observed period the intrinsic luminosity is inferred, and hence the distance at which such a star would appear as bright as it does.

The current, though not unanimously accepted, view is that cepheid variability is due to a radial pulsation of the star. The theory is highly developed, but has not yet accounted for the principal empirical relations.

Stellar Evolution

There is good reason to believe that the Galaxy was at one time composed of gas occupying approximately the same volume of space as the present stellar system. Quite well-understood processes must have produced local concentrations of density in the gas; these must then have contracted by self-gravitation to form stars. The foregoing outline of current views on stellar structure has been framed so as also to indicate the current view of the subsequent evolution of such stars.

This view is admittedly rudimentary. Besides leaving some important problems unsolved, it provokes fresh problems. For instance, its application leads to the conclusion that many stars are about 5,000 million years old. But the brightest existing stars would exhaust their proton-supply in a small fraction of this time. Consequently, they must have come into existence comparatively recently or else have been able to replenish themselves from outside. In either case, the essential process of stellar formation from interstellar matter must continue at all epochs.

Again, there are reasons for inferring that the original material of the Galaxy was pure hydrogen and that all other elements have been synthesized from this. But the existing relative abundance of these elements would require them to have been synthesized in conditions not to be found inside or outside any normal stars. The only known possible occurrence of such conditions is in a super-nova, i.e. a star having the outward characteristics of an ordinary nova but attaining a far higher maximum brightness. This possibility has the crucial advantage of providing also the means of dispersing the products of the synthesis. On the other hand, it makes the surprising demand that practically all the elements except hydrogen and helium in all existing stars must at some time have originated in other stars that became super-novae, and must subsequently have been acquired from interstellar space by the existing stars.

Whatever other merits these ideas may possess, they show that the problem of stellar evolution has now become that of the relation between the stars and matter in interstellar space. Even were we otherwise unacquainted with the existence of such matter, consideration of stellar evolution would demand it. Actually there is ample evidence for the existence of about as much interstellar as stellar matter, which, however, is so extremely attenuated that it has only fairly recently been discovered. It is the subject of much contemporary astrophysical investigation which lies outside the scope of this chapter. Henceforth, the problem of evolution in astronomy will be that jointly of the stars and interstellar matter. A particular problem in this connection is that of the occurrence of double stars. Probably about one half of all stars are members of binary (or more complex) systems. It now seems that this circumstance can be explained only by ascribing an essential role to interstellar matter, but the processes involved are still obscure.

Historical Review

Observational

In the past century, the greatest single development in the *detection of radiation* from celestial objects has been in the use of photography. The first successful experiments in celestial photography were made at Harvard in 1850, and the result of one of these, W. C. Bond's daguerreotype of the Moon, was shown at the Great Exhibition of 1851. Double-star photography was started at Harvard in 1857, but it was not until 1882, when a photograph of a comet at the Cape Observatory showed a rich background of stars, that the full possibilities of stellar photography were glimpsed. About 1858, Warren de la Rue (1815–89) obtained the first solar photographs having scientific value. Ever since these early days, there have been successive improvements in photographic emulsions and techniques.

Spectrum analysis was established as a means of identifying chemical elements by R. Bunsen and G. Kirchhoff in 1859 when the latter was also the first to use it to show the existence of known elements in the Sun. This was the second greatest advance in astronomical physics. The greatest was when Galileo and Newton showed that celestial and terrestrial bodies are subject to the same laws of motion and gravitation. Now it was shown that they are made of the same materials. Kirchhoff had discovered also the physical conditions required for the formation of spectral lines, thereby showing that spectrum analysis would reveal not only the composition of remote matter but something about its physical state as well. Knowledge of this relation was much extended later by J. N. Lockyer (1836–1920). Much the greater part of astrophysical observation has been based on this concept. So far as the stars are concerned, although it gives information only about the outermost layers, yet this leads to, or provides a check upon, all that we know of stellar constitution.

The chief pioneers of stellar spectrum analysis were W. Huggins (1824–1910) and A. Secchi. Huggins developed techniques and was the first to apply them to many particular problems. He and H. Draper introduced the photographic method of studying stellar spectra. Secchi was the first to seek a classification of stellar spectra and laid the foundations of the system of spectral classes in current use.

In 1913 H. N. Russell constructed the first luminosity-spectral class diagram and so discovered the empirical correlation between these characteristics.

The other fundamental correlation, between mass and luminosity, was foreshadowed by J. Halm in 1911 but first gained prominence in 1924 as a result of theoretical investigations by A. S. Eddington (1882–1944).

We need scarcely remind ourselves that spectrum analysis in its modern guise consists mainly in the measurement of the variation of *radiation-intensity* through some wave-length interval and that this is the subject of some of the latest technical developments.

Theoretical

Present knowledge of the constitution of the Sun's atmosphere has developed from Kirchhoff's interpretation (1861) of his own observations. The first steps towards knowledge of its internal constitution were taken soon afterwards, partly in attempts to explain the non-uniformity of solar rotation discovered by R. C. Carrington (1826–75). Notably through H. Faye's attempt (1865), it came to be accepted that the interior is gaseous—though until about 1924 it was not generally thought to be like a 'perfect' gas—and that the energy radiated from the boundary must be generated in the deep interior. This generation had been attributed by H. von Helmholtz in 1854 to the gravitational contraction of the Sun. This hypothesis, later elaborated by Lord Kelvin (1824–1907), held sway for about sixty years.

In 1870 J. Homer Lane made the first attempt to calculate the temperature distribution inside the Sun. His two hypotheses were that the material behaves like a 'perfect' gas and that the equilibrium is convective. These were used in important developments by A. Ritter between 1878 and 1883, and R. Emden in his famous *Gaskugeln* (1907).

The current conception of stellar constitution is due above all to the work done by Eddington between 1916 and 1926. His starting point was that there need be no difficulty in accepting Lane's first hypothesis in the case of giant stars, and later he was able to explain by atomic theory why it actually holds good also for main sequence stars. But, following a suggestion made in 1894 by R. A. Sampson (1866–1939), he discarded Lane's second hypothesis and showed that the greater part of a star must in general be in *radiative* equilibrium. He was able to employ the principles of such equilibrium developed by K. Schwarzschild in 1906 and the mathematical methods of Lane,

Ritter and Emden. His work was being done when the quantum theory of the atom was attaining its present form. He speedily grasped its significance for his own problems, making fundamental use of its application to the theory of ionization (as initiated by M. N. Saha and developed by R. H. Fowler (1889–1944) and E. A. Milne (1896–1950) for a quite different astrophysical problem) and to the theory of the absorption of radiation.

In 1936 T. G. Cowling (b. 1906) showed that a gaseous star in which the energy-generating process depends upon a high power of the temperature is rendered stable by the possession of a central region in convective equilibrium. The carbon-nitrogen cycle was fairly definitely established as the predominant energy-generating process by H. von Weizsäcker and by H. A. Bethe in 1938. Notable among many other contributors to the current conception of main-sequence stars are B. Strömgren, F. Hoyle (b. 1915) and R. A. Lyttleton (b. 1911). Hoyle has shown that the properties of the sequence as a whole are apparently best explicable on the hypothesis of a great preponderance of hydrogen in the composition; to him also are due many of the ideas concerning stellar evolution on which the preceding account is based.

The generally accepted model of a main-sequence star is essentially Cowling's, but this has density and temperature distributions not differing greatly from those computed by Eddington.

It was R. H. Fowler who showed in 1926 that the white dwarf problem could be solved by use of the quantum theory of degenerate matter; further developments concerning white dwarfs are due principally to E. C. Stoner (b. 1899) and S. Chandrasekhar.

Outlook

To us it may well appear that a century ago astronomers were faced with a greater number of fundamental problems obviously requiring solution than we are today. But the problems were far less obvious to them at the time than they seem to us in retrospect. Also the solutions were achieved by observational methods and results and physical theories then undreamt of.

Therefore we must not expect to foresee future developments: we can only note current trends. There is reason to believe that known outstanding problems of stellar constitution should be solved in the next few years. Already the study of interstellar matter is gaining comparable importance. It may be expected to elucidate questions of stellar evolution and possibly of stellar magnetism, of cosmic rays, and of sources in the Galaxy of radiation in radio frequencies (though these sources may be stars of some sort not otherwise known to astronomy). All such work may be expected to lead naturally to the study of the constitution of the galaxies in general.

It is even more difficult to forecast observational developments. Photo-electric devices are of growing importance; observations have already been got from rockets; radio astronomy is a rapidly expanding subject. These indicate some of the trends.

As to physical theory, fundamental developments may be expected in nuclear physics and the theory of 'elementary' particles. These are bound to have profound consequences for astrophysics. The recently advanced hypothesis of the continuing creation of matter is but one indication that, whereas such understanding as we possess of the constitution of the stars depends upon our knowledge of certain atomic processes, our understanding of still greater astronomical systems must depend upon a knowledge of even more fundamental atomic phenomena.

It is not unfair to say that while, a century ago, the internal constitution of a star was completely unknown and the idea of its evolution scarcely conceived, we have today a well-established body of knowledge on both subjects which can be used with some confidence in approaching the large problem of the structure and evolution of the universe.

CHAPTER X

THE STRUCTURE OF THE UNIVERSE

by SIR HAROLD SPENCER JONES, F.R.S.

UNTIL the time of William Herschel (1738–1822) astronomers had concerned themselves mainly with the positions and motions of the heavenly bodies and not with their distribution in space. This is the more surprising because it is at once apparent, when we look at the sky with the naked eye, that the distribution of the stars on the celestial sphere is far from uniform. If the stars were uniformly scattered throughout the sidereal universe, the relative numbers of stars seen in two different directions would be a measure also of the relative extent of the universe in those directions. On the other hand the inequality of distribution on the sky might be due to a real inequality of distribution in space. Knowledge of the distances of the stars is essential before it is possible to distinguish between the two alternatives. Such knowledge was not available in Herschel's day, for it was not until 1838 that the distance of a star was first measured.

Herschel therefore made the assumption that the stars are uniformly distributed. With his 20-foot telescope he counted the number of stars visible in the field of view (one-quarter of the area of the full moon) in a large number of regions distributed over the sky. In some regions he counted several hundred stars, in other regions only one or two. He came to the conclusion, which had previously been reached on more general grounds by Thomas Wright of Durham (1711–86), that the sidereal universe is a flattened system, shaped roughly like a disk or grandstone, whose central plane is defined by the Milky Way. Later in his life he realized that the assumption of uniform distribution was a wider departure from the truth than he had at first supposed, for he found many closely compressed clusterings of stars which could only be explained as real physical clusters.

Island Universes

The idea that our sidereal universe is one of many separate universes—island universes in space—which also had been suggested by Wright, was independently proposed by Herschel. He interpreted the nebulae as swarms of stars which his telescope was unable to resolve; they looked in his telescope much as the Milky Way looked to the naked eye. If this were so, they must be very remote systems; he even supposed that some of these distant systems were comparable in size with our own. He subsequently modified his views somewhat, for his detection of nebulous stars, surrounded with a 'shining fluid, of a nature totally unknown to us' showed that in our sidereal system there was matter which could not be resolved into stars. Moreover the nebulae were not uniformly distributed over the sky; they were most frequent near the poles of the Milky Way, which seemed to indicate a relationship to our system. Nevertheless, Herschel up to the end of his life considered that not all the nebulae were in our sidereal universe but that some were, in fact, external universes.

In 1845 and the next few years, Lord Rosse (1800–67) with his 6-foot reflecting telescope,[1] having a greater light-grasp and resolving power than Herschel's largest telescope, showed that the nebulae were of two types. Some showed a definite spiral structure and could apparently be resolved into separate clusterings of stars; others showed no definite structure and no stars could be seen in them. It might be supposed that, with greater resolution, these also would be resolved into discrete groupings of stars. But in 1864, Sir William Huggins (1824–1910) obtained the first spectrum of a nebula and found it to consist of bright lines, the type of spectrum which is given by a glowing gas. In the years that followed he observed the spectra of many other nebulae, and found that many of them also had bright line spectra. Herschel's suggestion of a 'shining fluid' was thereby confirmed. The nebulae which could be resolved into separate clusterings of stars as well as some others not then so resolvable gave continuous spectra, such as might be expected from distant clusters of stars.

[1]i.e. A telescope with an aperture of 6 feet diameter, according to the conventional designation. Herschel was exceptional in describing his telescopes (e.g. the 20-foot) by focal length instead of aperture.

Astronomical Photography

Meanwhile a new tool was being developed which was destined to have far-reaching effects in astronomy. In 1850 the first successful astronomical photograph—a daguerreotype of the moon—was obtained by W. C. Bond (1789–1859) at Harvard. In that same year the collodion wet plate was invented, enabling photographs to be obtained with much shorter exposures and with a technique that was not too difficult. Our present knowledge of the structure of the Universe is based almost entirely upon photographic observations. The photographic plate can store up faint impressions and reveal what it is impossible for the eye to see, providing at the same time a permanent record and saving much time at the telescope.

Little advance upon the speculative ideas of Herschel about the structure of the Universe was possible until knowledge of the distances of the stars could be obtained. This was slow in coming, for the determination of the distance of a star depends upon the measurement of the displacements in the star's apparent position as the earth moves round the sun. The angles to be measured are so small that extreme accuracy is required. It is difficult to obtain such accuracy by visual observations, which are moreover tedious and slow. The successful application of photography to this problem depended upon the development of special techniques and was not achieved until the early years of the present century. By the year 1900, not more than a score of stellar distances had been reasonably well determined, but so much effort has been devoted to these measurements in recent years that the distances of several thousand stars are now known with high accuracy.

Interstellar Matter

The problem of investigating the structure of the universe has been complicated by the discovery that obscuring matter is widely distributed throughout the sidereal universe. In 1895 E. E. Barnard (1857–1923) obtained excellent photographs of the Milky Way and drew attention to many dark markings or regions devoid of stars, occurring often in some of the densest portions. Herschel had noticed how often the sky appeared to be swept clear of stars in the immediate vicinity of a nebula; Barnard's dark markings were often, but not invariably, associated with bright nebulosity. The natural interpretation of such blank patches would be that they are devoid of stars: but this would imply that there were many long vacant lanes pointing directly towards us. Barnard's studies of the Milky Way showed that there were far too many of these blank patches for this interpretation to be tenable. The only plausible explanation is that the regions appear devoid of stars because there is obscuring matter between us and the stars which, like a fog or a cloud of dust, prevents us from seeing them.

This explanation has been well established. Moreover, the interstellar regions are not empty but contain matter in an extremely tenuous state. In the spectra of many distant stars of early spectral types, sharp absorption lines, produced by absorption of sodium and calcium, are observed, which do not share in the Doppler displacement caused by the star's own motion (a Doppler displacement is a shift of the spectrum lines—towards the red end of the spectrum if the star is receding and towards the violet end if it is approaching); these absorptions are produced by the interstellar gas. The gaseous nebulae can be thought of as being condensations in this interstellar gas, and they shine because they scatter the light of stars embedded in them. Associated with the gaseous matter, there is also a considerable amount of matter in the form of dust, which has a very strong obscuring effect. The dust and the gaseous matter are normally both present in the condensations and so it is frequently found that the bright nebulosity and the dark obscured patches are very much intermingled with each other.

This interstellar matter is very strongly concentrated towards the plane of the Milky Way and limits the distance to which it is possible to see in or near that plane. Away from the plane of the Milky Way, as the light from a distant star traverses a smaller length through the absorbing medium, the obscuration becomes less, although it is not negligible even when looking in a direction at right angles to the Milky Way. The existence of the obscuration explains why, before its existence and importance had been realized, various investigations of the structure of the sidereal universe always placed the Sun in a central position. If you are in a forest in a mist and at a distance from the edge of the forest which is greater than the distance to which it is possible to see, you might conclude that you are in the centre of the forest, although you may be quite near its edge.

Stellar Distances

The knowledge of stellar distances that can be obtained from direct measurement is limited to the stars within a distance of about 250 light-years; for a star at a greater distance, the parallactic angle is smaller than the probable error of its measurement. Practically speaking, we can determine by direct measurement the distances of those stars only which are comparatively near the Sun. But the scale of distances can be extended by the fact that it is possible to derive the intrinsic luminosity of a star from certain characteristics in its spectrum. Two stars of the same surface temperature, which differ greatly in intrinsic brightness, have almost identical spectra, because the nature of the spectrum is determined primarily by the temperature. But there are certain features in the spectrum which are sensitive to luminosity. By using these features, the stars with directly measured distances providing the basis for the correlation, the intrinsic luminosity of a star can be determined; the apparent brightness of the star being known, the distance readily follows. This indirect method of deriving distances enables information about the distances of stars out to a few thousand light-years to be obtained. Even such distances are quite inadequate for obtaining any idea of the size of the sidereal universe.

The key to the solution of the problem was provided by a special class of variable star known as the cepheid variables, the name being derived from the first star of this type to be known, Delta Cephei. These stars have a characteristic form of light variation, the rise from minimum to maximum being more rapid than the fall from maximum to minimum. Their periods are rather short, mostly ranging from a few hours to a few weeks. The variations in brightness of a cepheid variable are perfectly regular, and the period is remarkably constant. It has been found that these stars are in a state of regular pulsation and that this is the cause of the variations of brightness.

In 1912 Miss H. S. Leavitt (1868–1921), who was studying at the Harvard Observatory a large number of these stars in the smaller Magellanic Cloud, found that there was a very strong correlation between the period of light variation and the mean apparent brightness of the star. The Magellanic Cloud is a distant system, outside our Galaxy, which is visible because it is not in the Milky Way.

Therefore it was justifiable to assume that all the stars in the Cloud are at essentially the same distance from the Sun. The interpretation of Miss Leavitt's result was, consequently, that there is a strong correlation between the period of a cepheid variable and its mean intrinsic brightness. By using those cepheid variables whose distances have been directly determined, this relationship can be standardized so as to provide a definite relationship between period and intrinsic luminosity. In other words, if the period of any cepheid variable is determined by observation, its intrinsic brightness and its distance can be inferred.

The importance of this result is due to the fact that the cepheid variables are all stars of very high intrinsic luminosity and they are therefore visible to great distances. They serve as standard beacons to the astronomer, and enable him to derive reliable information about distances which are far beyond the range of direct measurement.

In 1914 H. Shapley (b. 1885) at Mount Wilson commenced the study of the special group of objects known as *globular clusters*—globular aggregations of thousands of stars, strongly concentrated towards their centres. About one hundred of these objects are known, none of them being in the Milky Way; with a few exceptions they are all in one hemisphere of the sky. In all of them cepheid variables are found to occur, and it thus becomes possible to derive the distance of each cluster. They are found to be distant systems, their distances ranging from about 20,000 to more than 100,000 light-years.

Dimensions of the Galaxy

The interpretation of the observations is that the globular clusters are members of the Milky Way system and that they are co-extensive with that system. They are found mainly in one half of the sky because the sun occupies a very eccentric position in that system. None of them are to be seen in the Milky Way because the obscuring matter hides them from our sight, but those which are at considerable distances from the Milky Way can be seen because the light from them travels through a relatively short path of the obscuring matter before reaching us. The clusters outline a more or less spherical region, with a diameter of about 100,000 light-years, the Sun being at a distance of about 30,000 light-years from the centre. Thus the globular clusters and the cepheid variables together have provided the clue to the approximate dimensions of our sidereal universe and to the position of the Sun in that system. Cepheid variables are found also in many of the star-clouds of the Milky Way, and distances have been found for these clouds of from 10,000 to 20,000 light-years. Because the Sun lies almost in the central plane of the system, in which the obscuration is strongest, it is not possible to see the stars of the Milky Way out to its greatest depths.

Galactic Rotation

The strongly flattened form of the Galaxy suggests that it is in rotation; rotation would seem, indeed, to be essential for stability. A general theory of galactic rotation was developed by B. Lindblad (b. 1895) in 1926. Shortly afterwards J. H. Oort (b. 1900), from the study of the motions of the stars in the near vicinity of the Sun, found conclusive evidence of the rotation. If the Galaxy is in rotation, it will not rotate like a solid body, in which the linear speed is greater, the greater the distance from the centre of rotation. The rotation will be in accordance with the general laws of motion under gravitational attraction, and as in the solar system, the greater the distance from the centre the slower the speed. From the investigation both of the line-of-sight motions of the stars and of their proper motions across the line of sight, it was found that the stars on one side of the Sun were moving faster than the Sun and that those on the other side were moving slower. Interpreting this as the effect of rotation, the direction to the centre of the rotation is the direction towards the constellation of Sagittarius, the brightest and densest portion of the Milky Way. This is also the direction to the centre of the system outlined by the globular clusters.

Mass of the Galaxy

The velocity of the orbital motion of the stars in the vicinity of the Sun is about 150 miles a second, but the distance to the centre is so great that about 200 million years are required for one revolution to be completed. By the application of dynamical principles, the controlling mass can be estimated and is found to be about one hundred thousand million times the mass of the Sun. From the estimates which can be made of the density of interstellar matter, it has been concluded that the total mass of the interstellar matter in the Galactic System is about equal to the total mass of the stars. About half of the matter in the system has therefore condensed into stars, the other half still existing in its primitive state of gas and dust particles. The stars, as they move through space, are continually drawing in this matter by their gravitational attraction and so are gradually sweeping space clean, but the average distance between the stars being three or four light-years, the sweeping process is a very slow one.

Existence of Island Universes

The question whether there are island universes in space, beyond the bounds of the Galactic System, as Herschel had believed, was still unsettled thirty years ago. The avoidance by the spiral nebulae of the Milky Way seemed to indicate a relationship to the Galaxy and to argue against their being island universes. The strong obscuring effect of the dusty matter in the Milky Way regions, which satisfactorily accounts for none of the spiral nebulae being visible in the Milky Way, was not realized at that time. The answer to this question could not be obtained until a telescope of greater light-grasp and greater resolving power than any then available could be brought to bear upon it. The completion, in 1918, of the 100-inch reflector at the Mount Wilson Observatory opened the way, for its light-gathering power was nearly three times greater than that of the largest existing telescope.

The brightest of the spiral nebulae, the great nebula in Andromeda, which is faintly visible to the naked eye, is the one which can be studied in the greatest detail. The photographs of the Andromeda nebula showed it to be a system with many strong resemblances to the Galaxy. It is a flat system, seen at an oblique angle, containing aggregations of stars which are very much like the star-clouds of the Milky Way; there are many patches of bright nebulosity and, intermingled with these, are many dark patches caused by obscuring clouds. Discrete stars can be seen and, when photographs taken at different epochs are compared, it is found that many of the stars are variable in brightness. A number of these variable stars proved, upon further investigation, to be cepheid variables. Their periods were determined and their intrinsic luminosities inferred from the period-luminosity relationship; the distance of the Andromeda Nebula was found, in that way, to be about 750,000 light-years. Such a distance places it well beyond the limits of the Galaxy and indicates that it is indeed an island universe. From time to time also, a nova or new star was seen suddenly to flare up in the system and then gradually to fade away. From the study of novae in our own Galaxy it is known that, when at maximum brightness, a nova (unless of the very exceptional class of supernovae) has a luminosity about ten stellar magnitudes brighter than the Sun. Assuming that the novae observed in the Andromeda nebula are of a similar intrinsic brightness at maximum, the distance obtained for the system is in accordance with the distance derived from the cepheid variables.

In several other of the brighter and nearer spiral systems cepheid variables and novae were found and it was possible to determine their distances. They proved to be further away than the Andromeda nebula and to be, therefore, also island universes. The question whether these objects were part of the Galaxy or were separate systems was at last settled. The reason why none of them are seen in low galactic latitudes is that they are hidden by the great depth of obscuring matter. These distant systems, lying outside our Galactic system, are usually termed *extragalactic nebulae*, to distinguish them from the gaseous nebulae in the Milky Way; they are also often referred to as external galaxies or merely as galaxies, to distinguish them from our Galactic system or Galaxy.

The extragalactic nebulae are seen at all inclinations to the line of sight. When seen edgewise-on they are found to be highly flattened systems, with a belt of obscuring matter along their median plane, analogous to the marked concentration of the obscuring matter in the Galaxy into the plane of the Milky Way. When seen broadside-on, most of them show the characteristic spiral structure, suggesting that, if the Sun were not in the plane of the Milky Way, our Galaxy would appear to us as a spiral system. Some of them show a great deal of structure in the form of star clouds and even discrete stars; others show little structure, as though they are in a less advanced stage of evolution, most of the matter not yet having condensed into stars.

The line-of-sight velocities of many of the external galaxies have

L

been determined. In the case of the nearer systems which are seen edgewise or nearly edgewise, it is found, by measuring the line-of-sight velocity at different points of the system, that they are in rotation, the rotational velocity decreasing outwards from the centre. It becomes possible, therefore, to make an estimate of the controlling mass, which proves to be of the same order as that of our Galaxy. The periods of rotation are also comparable with those of the Galaxy.

Dimensions and Distances of External Galaxies

The question remains whether the dimensions of these systems are comparable to those of the Galaxy. When the distances have been determined and the angular diameters measured from photographs, the actual linear diameters can be inferred. It appeared at first that their dimensions are of much the same order but that our Galaxy was appreciably larger than any of the others. Further investigations have resolved this difference; in the first place, the dimensions of our Galaxy were initially overestimated, because the effects of the dimming of the light of distant stars by the obscuring matter had not been sufficiently realized; in the second place, it is difficult from photographs to assign correctly the faint outer limits of a spiral system and, when observations were made with a sensitive photo-electric cell, the limits of the Andromeda nebula were found to be well outside those which had been assigned from photographs. It now appears that the Andromeda nebula is a system of about the same size as the Galaxy, and that, by and large, the galaxies are much of a size.

By assuming, as a working hypothesis, that the galaxies are of about the same size and that they contain about the same amount of matter and are of about the same intrinsic brightness, the way is open to make estimates of the distances of systems which are too far away for individual stars to be seen in them. Each of the two assumptions, uniformity in size and uniformity in brightness, provides an estimate of distance; the two estimates are found to be reasonably concordant and the mean of the two can be adopted as a provisional value.

The first measurement of the line-of-sight velocity of an extra-galactic nebula was made in 1912 by V. M. Slipher (b. 1875) at the

Lowell Observatory. By the year 1925 he had determined the line-of-sight velocities of forty-one systems. For the distant and fainter systems he used a very fast, short-focus camera with a small-dispersion spectrograph. The accumulation of velocities with the aid of a 24-inch refracting telescope was slow and laborious, for with the exception of the few relatively near and bright objects long exposures are required. The Andromeda nebula was the first to be observed and it was found to be approaching the earth with a speed of about 190 miles a second. But as further observations were secured, it was found that almost without exception the extragalactic systems are receding from the earth, and that the fainter they appear, the more rapid are their velocities of recession. The velocities are very much greater than those of the stars in the Galaxy.

The Velocity-Distance Relation

When the results were analysed statistically, it was found that the motion of the sun with respect to the external galaxies is about 175 miles a second in the general direction of the bright star Vega. This is closely equal to the velocity of the Sun round the centre of the Galaxy, so that the motion of the Galaxy amongst the other systems must be small. The velocities of the external galaxies were found to be closely proportional to their distances, the residuals from this relationship being only about 100 miles a second—which could be attributed in part to errors in the assumed distances and in part to the proper motions of the individual systems. The velocity of approach of the Andromeda nebula shows that the galaxies do have their own peculiar velocities. As the velocities of recession increase, on the average, at a rate of roughly 100 miles a second per million light-years of distance, the peculiar motions become relatively unimportant at distances exceeding ten million light-years or so; galaxies at such distances must be considered relatively near.

The velocity-distance relation serves a double purpose. In the first place, it enables distances to be assigned to the galaxies which are more accurate than those based upon the assumptions of uniformity of size and of intrinsic brightness; in the second place, it provides a means whereby the distances of very remote faint systems can be determined with very considerable accuracy. It is found that the spectra obtained by giving long-exposures on faint objects with a small dispersion spectrograph show very clearly the strong *H* and *K* lines of ionized calcium; from the wave-lengths of these lines, using the spectrum of a comparison source, such as the iron arc or neon vapour, as a basis of reference for measurement, the line-of-sight velocity can be readily and accurately derived. M. L. Humason (b. 1891), using the 100-inch reflector at Mount Wilson, gave exposures up to 30 hours to obtain spectra of faint distant extra-galactic nebulae in order to determine their velocities. It is not possible even with the 100-inch telescope to see these faint objects and it is necessary to select some adjacent star and to guide on it during the exposure, setting it in the field of view so that the invisible nebula is on the slit of the spectrograph. The most distant galaxy whose velocity has been measured is at a distance of about 240 million light-years and its velocity of recession is 26,000 miles a second, about one-seventh of the velocity of light.

The More Distant Galaxies

Long-exposure photographs show many galaxies that are still more distant. Near the galactic poles, where the effects of obscuration by interstellar matter are least, the photographs show more external galaxies than stars in our Galaxy. These distant island universes can be recognized by their diffuse images, which are different from the sharp images of the faint galactic stars. By extrapolation, based on the limiting magnitude recorded on the plates, it is estimated that the most distant objects shown on the photographs are at a distance of about 500 million light-years. The velocities of recession of these remote systems have not yet been measured but may be expected to be about 53,000 miles a second. With the 200-inch reflector at Mount Palomar, it will be possible to photograph their spectra and to determine their velocities, and also to survey space to a distance of about 1,000 million light-years.

The observed density of the external galaxies per square degree of area of the sky falls off progressively towards the galactic plane, because of the increasing amount of absorption by the interstellar matter. Allowing for the effects of this absorption, it is estimated that within a distance of 500 million light-years, the effective limit which can be reached with the 100-inch reflector, there are about 100 million separate galaxies.

Apart from some clustering of the extragalactic nebulae (which, however, is considerable) their distribution in space seems to be more or less uniform. The average distance between a nebula and its nearest neighbour in space is about two million light-years. Out to the limits of observation there is no evidence that the density of the galaxies in space is decreasing, so that it is probable that the observable region is small compared with the region which remains beyond. With the 200-inch reflector it will be possible to survey space to twice the distance that is possible with the 100-inch reflector, and it will be of interest to find out whether the space-density of the galaxies shows any sign of decreasing at this greater distance.

The Overall Picture

The picture of the Universe which has been revealed by observations with modern large reflecting telescopes is therefore somewhat as follows: space, out to the present extreme limits of observation, is populated with a vast number of 'island universes'; these universes are more or less uniformly distributed, except for some local clustering here and there; they are flattened disk-like structures, in slow rotation and held together by their mutual gravitation; they are, by and large, pretty uniform in size, in intrinsic brightness, and in mass, each containing enough matter to make about 100,000 million stars of mass equal to that of the sun; the state of aggregation into stars varies appreciably from one system to another, but even where it has proceeded furthest there remains a considerable amount of matter in the form of diffuse gas and dust particles. Within the distance of 500 million light-years, which is the limit of present surveys, there are about 100 million separate island universes.

It is well to ponder for a moment over this distance of 500 million light-years, for it is so vast that the mind does not readily comprehend it. It can be better conceived if we think of it in this way. When the light by which these remote systems are photographed set out on its long journey through space, the face of the earth was very different from what it is now. Many and great changes have occurred while the light has been travelling on through space. Mountain ranges have been thrown up as the earth's crust has crumpled and have been gradually worn down by the action of rain, running water, moving glaciers and blown sand, forming sedimentary deposits beneath the waters on the globe. New mountain ranges have in turn been thrown up and worn down, and this process has been repeated many times. Dinosaurs, flying reptiles, and other animals long since extinct have appeared on the earth, have passed through their course of evolution and have disappeared, giving place in turn to new and different forms of life. It was not until the light was in the very last lap of its journey that man himself appeared upon the earth, and it was passing amongst the stars in our immediate vicinity before a telescope was constructed which had a sufficient light-grasp to reveal the vast universe from which the light had come as a very faint speck on a photographic plate.

The Expansion of the Universe

The observational fact that the separate island universes are receding from us with velocities that are proportional to their distances is what we should actually observe if space as a whole were expanding, carrying all the universes with it. In recent years, theories of cosmology have formulated various models of the Universe which satisfy the mathematical requirements of generalized relativity, according to which the geometry of space is determined by the contents of space. A. Einstein (b. 1879) and W. de Sitter (1872–1934) obtained solutions, representing universes that are static, homogeneous and isotropic. De Sitter's universe required that there should be an apparent recession of remote objects, due to the slowing down of atomic vibrations; this mathematical result was obtained before the observational fact of the displacements to the red of the lines in the spectra of the external galaxies had been established. When observations showed that there was, in fact, such an effect, it seemed to give strong support to de Sitter's universe. This universe is, however, a completely empty universe; if it contains any matter it is not possible for it to remain static or, in other words, changeless. Einstein's universe, on the other hand, can contain matter, but will not account for the phenomenon of the red-shifts of the spectral lines.

In 1927 G. Lemaître (b. 1894) proved that a static universe was not the only kind of universe satisfying relativity requirements, and in 1930 Eddington (1882–1944) showed that such a universe would in fact be unstable; the Universe must therefore either expand or contract. If we interpret the red-displacements in the spectra of the galaxies as indicating line-of-sight motions, and not as evidence of a slowing down of atomic vibrations, we conclude that the separate universes are all receding from us and necessarily, of course, all receding from each other. The expanding universe of Lemaître is something more than this; there is not merely expansion of the material system of the galaxies—there is also an expansion of the closed space in which they exist. If we attempt to go backwards in time, to find out what was the state of the Universe before the expansion began and what caused the expansion to start, we find ourselves in the realms of conjecture and speculation; there is no certainty and one person's guess is as good as another's.

Latest Knowledge

The past century—and, indeed, the past forty years—have seen a revolution in our knowledge of the structure of the Universe. When the first primitive ideas about its structure had been formulated, nothing was known about its dimensions, for the distances of only a very few of the nearest stars had been measured, and those not very accurately. By a sort of intuition it was thought that there might be island universes in space, but this was merely intelligent speculation. With the equipment and techniques available to the astronomers of a century ago it was, indeed, impossible to derive a true picture of the Universe.

Our present knowledge has been made possible through the construction of large telescopes with great light-gathering power and through the invaluable help afforded by the spectroscope and the photographic emulsion. These alone have not sufficed; the human element has been all-important—the long night watches at the telescope, often exposing for several hours on several separate nights to obtain a photograph of a single faint object or of its spectrum; the patient scrutiny of the plates, and the comparisons between successive photographs of the same object, to discover variations in the brightness of individual stars or other changes that may have occurred; the gradual building-up of methods of probing with confidence deeper and deeper into space; and the co-ordination, synthesis and discussion of all the data, obtained in various ways. It was said of Herschel that he broke the barriers of the Heavens;

the barriers have now been thrown wide apart and the astronomical horizon has been pushed back to distances that were undreamt of in Herschel's time. Our present knowledge of the structure of the Universe is, nevertheless, far from complete and we can confidently expect that in the course of the next century, with still larger instruments and possibly with new techniques, much fuller information will be obtained.

(*Above*) Two illustrations of the spiral nebula in Canes Venatici (the upper from a drawing made with Lord Rosse's 6-ft. reflector, taken from the *Philosophical Transactions of the Royal Society*, 1850) and the lower from a photograph taken with the 60-in. reflector of the Mount Wilson Observatory.

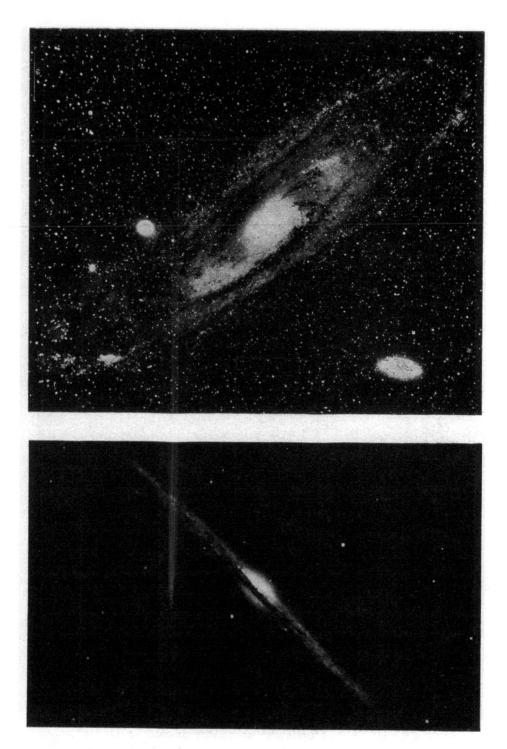

(*Above*) The Great Nebula in Andromeda, the nearest of the spiral systems, photographed with the 24-in. reflector of the Yerkes Observatory.

(*Below*) The spiral nebula in Coma Berenices, a galaxy seen edgewise-on, photographed with the 60-in. reflector of the Mount Wilson Observatory.

CHAPTER XI

ORGANIC EVOLUTION

by E. B. FORD, M.A., D.SC., F.R.S.

RARELY, a few times in a century at most, there dawns a new phase in human thought. Never in the biological sciences has the change been so sudden and illuminating as that produced when a joint statement by Charles Darwin and Alfred Wallace was read to the Linnean Society on 1st July, 1858. On that occasion they suggested for the first time a mechanism, that of natural selection, which could reasonably be held responsible for organic evolution. This Darwin expanded the following year in the greatest of his books, the *Origin of Species* (published on 24th November, 1859). Here he went much further, not only showing in more detail how evolution could be brought about, but supplying a vast body of evidence to demonstrate that it had actually taken place.

The evolutionary concept, that one form of life has changed into another so that distinct organisms are not separately created but descend from a common ancestor, is an ancient one, foreshadowed by the Greek philosophers and naturalists. In the middle of the eighteenth century it received a new impetus from an unexpected source. This was the work of Carl Linné, generally known as Linnaeus:[1] an impetus because it resulted in the first rational classification of organisms and indicated the fundamental similarities of certain clearly defined groups; an unexpected one, since Linnaeus himself believed in special creation and actually undertook his studies to unravel the multitudinous separate acts of the Creator. Yet his great contributions to systematics certainly encouraged evolutionary speculations, such as those of Buffon in the latter half of the eighteenth and of Lamarck at the beginning of the nineteenth

---

[1]Published in his books *Systema Naturae* (first edition, 1735) and *Species Plantarum* (1753). The tenth edition of *Systema Naturae* (1758) is taken as the starting point for all zoological (though not for all botanical) nomenclature.

centuries; however, their theories, false in themselves, carried no widespread conviction at the time.

That was still the situation one hundred years ago. Evolution appeared a mere possibility, though little recognized as such; for no convincing body of evidence had been collected to demonstrate it, nor was the mechanism actually responsible for it even suspected. Ninety years ago all this had been changed: it remained only to convince scientific sceptics, and they were many, of the truth of Darwin's views and to elaborate and in important points to correct them. Thus the whole concept of organic evolution as an ascertained and understood fact falls within the century covered by this volume: the period 1851 to 1951.

Any general account of this subject must cover two distinct topics: first, a survey of the *evidence* that evolution is responsible for the diversity of organisms which live today and have lived upon the earth; secondly, a discussion of the *mechanism* by which the process is brought about. As already indicated, both these aspects have been developed almost from their foundations since 1858, so that they must each be considered here. This is not a scientific article in the strict sense, in which each statement should be substantiated by exact references or by other means. But the facts which convinced the world of evolution are not mere corroborative evidence of that kind. They form the basis for conclusions which represent one of the triumphs of biological thought and research during the past hundred years and must therefore be reviewed before the working of the evolutionary processes is described.

The Evidence for Evolution

Proof that the structure of living organisms has been moulded by a process of evolution is derived from a number of distinct sources. Its cogency is, of course, greatly enhanced by such independent corroboration.

Palaeontology

The most direct evidence is that supplied by palaeontology, the study of fossils which, as is well known, represent the skeletal remains of animals and the supporting tissues of plants preserved in the rocks. Such relics are generally replaced by minerals but often so as to retain their shapes, sometimes with exactitude. Setting aside the igneous formations, due to volcanic eruptions, the different types of rock can usually be dated, at least relatively, independently of the fossils they may contain, since they have been formed in sequence upwards from the most ancient. Where the seams run more or less horizontally, as they do unless contorted, we can be sure that the upper of two rocks is the newer.

Now when we examine rocks of increasing age, it is found that the fossils embedded in them differ *progressively* from the forms alive today, and in two respects: the proportion of types that have left no survivors steadily increases, while the remainder differ from their modern representatives to an increasing degree. Moreover, in some instances, present-day forms, for instance the horse, can be traced in continuous sequence through older and older rocks to a creature that would never have been recognized as its ancestor had not the intervening stages been available for study.

The tangible evidence that evolution has occurred is thus decisive. It can be examined over a period of about 500 million years, back to the 'Cambrian' deposits. Those of earlier date are so affected by heat and pressure that the traces of fossils which they certainly contain are almost indecipherable. At the distant Cambrian epoch, our own group, the vertebrates, had already appeared but had progressed no further than fish of a primitive kind. We now know that from these were subsequently derived first amphibia (including the newts and frogs) and then reptiles, and that these gave rise independently to mammals and later to birds.

Geography

Turning to another aspect, the geographical distribution of animals strongly supports the concept of evolution. Darwin indeed largely convinced himself of it by this line of inquiry. It is not one easily summarized in a short space, but a few of its essential features may be mentioned here. Wallace had pointed out that most of the islands of the world can without uncertainty be placed in one of two groups: they may be *continental islands*, which were formerly part of a continent and have become separated from it by erosion or subsidence, or they may be *oceanic islands* which have formed at sea. The continental type will usually be situated on a continental shelf and contain sedimentary rocks (produced gradually by the accumulation of debris); such is Great Britain. Those of the oceanic group may or may not be on the continental shelf and will consist of coral or of volcanic rocks.

If the varied forms of life arose by special creation, there is no apparent reason why a consistent difference should exist between those indigenous to continental and to oceanic islands. If, on the other hand, evolution be a reality, only those capable of migrating or drifting through or over salt water could have colonized the oceanic type, though the continental would be subject to no such limitation; and that indeed is what we find. Furthermore, the extent to which the fauna and flora of continental islands differ from those of the neighbouring land-mass depends rather upon the length of time since their separation than on the breadth of the intervening sea. In general, it may be said that great differences exist between forms of life found upon areas long isolated from one another, even though they be close together. Thus the channel between Bali and Lombok in the East Indies is of remote antiquity, and the faunas of these two islands, no more than twenty miles apart, are fundamentally more dissimilar than are those of Portugal and Japan. Clearly these facts all constitute powerful evidence in favour of evolution.

Anatomy and Embryology

So, too, do those derived from comparative anatomy and embryology. Animals and plants can be arranged in groups whose anatomy is so similar as to preclude chance resemblance and strongly to suggest that they had a common origin. Thus the hearts and blood-vascular systems of all mammals are identical in their general plan; so are those of birds, but the two types are distinct from one another. Their design is, however, so combined in the reptiles as to form a basis from which both mammalian and avian systems could be derived and, as already indicated, this is in agreement with the fossil record.

Such examples might be multiplied indefinitely, but there is a special class of anatomical evidence to which reference must be made: the existence of vestigial structures. These are useless relics of organs functional in supposedly related creatures. The modern horse can be traced back to *Hyracotherium*, a creature the size of a large dog which lived about fifty million years ago. It retained most of the toes of an ordinary mammal: four on the front and three on the hind limb. These have now been reduced to one on each, but relics of two more persist today as thin functionless 'splint bones' attached to the skeleton of the feet. Similarly, traces of limbs can be found on dissecting a snake, and it is difficult to resist the conclusion that they are remnants of vanished legs.

Embryological studies are in a real sense a branch of anatomy. and they show that great structural changes occur during development; for organisms of fundamentally similar plan are more alike when young than when adult, suggesting that they have diverged from a common type. Those plants which are of simple construction and are thought to be *primitive* (to have departed relatively little from the common ancestor of all multicellular plants) pass through two distinct phases in their life-cycles. One, the *gametophyte*, is concerned with the production of sexual organs; the other, the *sporophyte*, with that of asexual spores. These two types alternate, one giving rise to the other. They are clearly distinct in the liverworts and mosses, the Bryophyta. However, the gametophyte phase becomes inconspicuous and short-lived in the ferns, the *Pteridophyta*, while it takes no part in forming the body of flowering plants, though even here a trace of it can still be detected in the reproductive organs.

Turning now to the vertebrates among animals, which exist only as fish and some other less specialized creatures at the start of the fossil record, we find clear indications of a fish-like stage in the development even of birds and mammals. For both these groups possess embryonic gills early in their development. Indeed for a short time they are actually open, and connect the throat with the surface of the neck. Furthermore, embryology sometimes reveals that certain parts of the body have had surprising histories: for instance, that the pineal gland on the upper surface of the human and other vertebrate brains is a relic of a third eye on the top of the head, still functional in some reptiles (and in the lampreys).

Unfortunately it is not possible here to elaborate beyond these few instances the great evolutionary significance of anatomy and embryology. It becomes indeed necessary at this point to say a few words of caution.

Classification

It is sometimes stated that classification provides important evidence for evolution. But modern classification consists in an attempt to reveal relationship: the systematist clearly appreciates that he is not grouping together those forms possessing organs of corresponding function, nor those with similar habits, but those which he believes, at the classificatory level he is concerned with, to be most nearly related. If then we survey the result and claim that the hierarchical classification of animals and plants so obtained supports the concept of evolution, we argue in a circle. A classification can only be appealed to thus if its accuracy has been ascertained independently of the facts used in constructing it. Opportunities of this kind exist but they are infrequent. I exemplify them from my own observations.

Being at one time interested in the significance of classification, I decided if possible to test its success in a group with which I was then familiar, the *Lepidoptera* (butterflies and moths). I found that their systematic arrangement had been reached, as it should be, by a study of a wide variety of features, anatomical and others, at all stages of their life-histories, but I was struck by the fact that one possible criterion had been omitted: that of chemistry. Indeed the chemistry of the varied pigments of these insects was not at that time sufficiently well known for the purpose. I therefore undertook a chemical study of them and ascertained their nature in a large number of butterflies and in some moths. Having done so, I applied the results to classification, and found that the various chemical types were not distributed at random among the different groups but were strictly related to the accepted arrangement, made in ignorance of them. That is to say, they provided independent evidence for the accuracy of a classification constructed with the avowed object of demonstrating relationship and the effects of evolution.

Observed instances of Evolution

It remains briefly to mention that the evolutionary processes can be reproduced artificially. This, of course, does not prove that they have in fact operated in nature: even so, such knowledge is really essential, for we should rightly feel suspicious if asked to accept a theory which could not be verified experimentally, even though well substantiated in other ways. In this connection it should, of course, be stressed that the profound effects upon animals and plants produced by domestication and cultivation are a clear indication that species can be permanently modified and made to evolve. For these alterations are not a mere response to environment, but persist after the agencies producing them have ceased to act. Here it is possible only to state that many types of evolutionary change, including the formation of new species, have been produced in the laboratory.

Finally, if it be questioned whether evolution has taken place in nature, it is worth mentioning that it has been observed to do so. Thus many different species of moths have completely altered their appearance in the industrial areas of England and other countries during the last hundred years, having lost their original pale or varied coloration and become blackish; an event which will be discussed more fully on pp. 178-9.

The main lines of evidence demonstrating that evolution is responsible for the diversity of life upon the Earth have now been indicated. Though examples have been reduced to a minimum, it will be realized that the proof they supply is overwhelming. At this point, therefore, it is possible to turn to the other aspect of the problem, and to consider by what means the process has been brought about.

The Methods of Evolution

Darwin and Wallace made one outstanding contribution to scientific thought: the theory of Natural Selection. This has often been summarized as the 'survival of the fittest', but even that phrase requires a brief explanation before its significance can be discussed.

Those individuals of any plant or animal species which are the best suited to their environment tend to be the most successful in reproduction, and so to contribute more to posterity than their less fortunate rivals. Consequently, if their superior qualities are to some degree inherited, these will in the course of generations be spread through the population, causing it to change or evolve.

Natural and Artificial Selection

The *natural* selection which produces this result is comparable with the *artificial* selection practised so successfully by Man. For gardeners and farmers are accustomed to choose for breeding purposes those individuals in which the qualities they prize are the most pronounced and, by this means, they have transformed wild plants and animals to suit their needs.

The importance of the selective process in nature was particularly impressed upon Darwin and Wallace by a consideration of variability and excess reproduction. Darwin, in particular, had studied variation, and he realized that it may affect every quality of the individual; and that even members of the same animal or plant species when living together, whether in a state of nature or under domestication, are very far from identical. Moreover, he particularly stressed in his writings the wastage which occurs during the life-histories of all organisms. If the numbers of a population are to remain constant, which they usually do when judged over a considerable period, whether they are subject to recurrent fluctuations or not, only two of the offspring of any pair on the average survive to breed. Yet a single pair of fish may produce several million fertile eggs, and a similar vast excess is frequent in plants.

At the other extreme, the death-rate before maturity is occasionally quite small: as in those birds (the guillemot, *Uria aalge*, for instance) which lay but one egg in a season, though each adult lives on the average for more than two years. Consequently there is still effective destruction even here, since the survival of a slight excess only above that required for replacement, say 2.1 instead of 2 offspring per pair, would produce an immense increase in comparatively few generations from an evolutionary point of view.

As Darwin correctly appreciated, when a variable population is exposed to elimination, the survivors cannot be a random sample of it; and indeed even a minute advantage will be of value in such long-continued competition. But unless the qualities which lead to success are to some extent inherited, each generation will start with the same constitution as the last, however much certain types may be favoured: the reality of *selection* remains, but it can contribute nothing to evolution if the variation upon which it acts is purely environmental in origin.

Heredity and Variation

Thus we reach the stage at which genetics, the study of heredity and variation, impinges upon evolution, and it was here that Darwin entered the field of pure speculation, for the mechanism of heredity was unknown to him. Indeed the system he assumed was the very reverse of the one which actually operates, and such mistakes as he made, and the contradictions he recognized but failed to resolve, were nearly all the outcome of that one fundamental error.

The principles and terminology of genetics are explained in Chapter XIV. A knowledge of them will therefore be presumed here. Nevertheless, it is necessary briefly to restate the essential features of heredity from an evolutionary point of view in order to apply them in this account.

Genes

Units of some sort, responsible in some kind of way for the characteristics of the organism, must be transmitted from parent to offspring if heredity has any physical basis at all. These units are known as *genes*, and the majority of them occur with equal frequency in the two sexes. Their most fundamental quality is their extreme permanence; for, though their effects on the organism depend on their interaction with one another and with the external environment (see Chapter XIV), they do not contaminate one another, or blend, when brought together into the same individual while, in addition, they have a high degree of intrinsic stability. That is to say, they are built up of molecules which, though large, are relatively stable in the chemical sense; in the language of genetics, the genes have exceedingly low mutation-rates.

Now a great proportion of them, though not all, are situated in the nuclei of the cells, where they are carried in the chromosomes, so that they exist in pairs (allelomorphs) the members of which are derived respectively from the two parents and obey the laws of Mendel. Consequently, those which arose far apart, in either sense of far-apartness, in time or in space, can be held long in reserve and ultimately brought together in new combinations some of which may prove of value to the individual.

It will be realized therefore that such 'particulate inheritance' possesses extraordinary advantages from an evolutionary point of

M

view, for it combines the possibilities of extreme heritable variability with extreme heritable stability. An immense array of forms can be presented for selection to work upon, while those which prove of advantage can be stabilized and preserved. We are, of course, here concerned with *genetic* variation, due to recombination and, to a very slight extent, to mutation, not with the environmental type produced by the different conditions in which the genes have to operate.

From an evolutionary point of view the genes may be classified as disadvantageous, neutral, or beneficial in their total effects on the reproduction or survival of the individual. Those which are disadvantageous cannot succeed in establishing themselves in more than a minute fraction of a population. They are constantly eliminated by selection and restored only by rare mutation (or by immigration from other environments in which they are a success), so that their frequency is maintained in equilibrium by these conflicting tendencies. Few genes are of neutral survival value, for this requires a rather accurate balance between advantage and disadvantage. Moreover, they spread at a rate which is calculable and very slow—so slow indeed that before the process has advanced far it is likely that conditions will have changed sufficiently to upset their neutrality. Thus it will be realized that if any gene is fairly frequent in a population, occurring in even a few per cent of available loci, it must have some advantage.

It should be noticed that the genes each affect the body in a variety of ways. Even those which produce visible characters of an apparently trivial kind—altering, for instance, the number of bristles on an insect or the colour of a flower—are usually responsible, in addition, for other less obvious but more important changes. If one of their effects chances to be useful it is unlikely that the others will be so too, for it is much easier to damage than to improve a highly organized system such as the body of an animal or plant. At the best therefore, the fate of a gene usually depends on whether or not its advantage outweighs its drawbacks.

These facts can be illustrated by a recent instance of evolution to which reference has already been made. Moths, like other organisms, have doubtless utilized all available genes which contribute to their bodily fitness. But some have not proved available; these, in addition to their valuable physiological action, produce excessive quantities of the black pigment *melanin*. This floods the coloured scales covering the wings and body so as to obscure the cryptic pattern upon which the safety of the insect depends; for such blackish or 'melanic' varieties fail to match the background of tree-trunks and lichen on which they are adapted to rest. Though actually hardier than normal specimens, they were consequently eliminated until man created an environment to suit them. This is the blackened countryside of industrial areas, where dark forms are no longer unduly conspicuous. Here then they have spread, and for the first time it has been possible for use to be made of the superior physiological qualities associated with them.

Considerations of the type just outlined are partly responsible for the fact that evolution often spreads non-adaptive or even harmful qualities through a species: for instance, genes useful during development may have an undesirable influence on the adult. Thus those speeding up growth may be favoured when the embryos compete with one another for space (as in mice, and many other mammals) or for food, even if they have injurious consequences later in life. It is worth noting that a premium may also be placed on the reverse process, that in which early development is retarded. This has been an important factor in human evolution, allowing a great extension of the baby phase with its period of learning: a result which could not have been achieved in the face of severe embryonic competition. The fact that Man and the great apes allied to him, on the average produce but one young at a birth has therefore had important consequences.

The Origin of Specific Differences

Selection produces more rapid evolutionary change when operating on multifactorial than on unifactorial qualities. The majority of those which separate related species are under multifactorial control, but the evolution of specific differences involves too many adjustments to be accomplished quickly, save in special circumstances. The process has always aroused interest, partly from its real importance and partly on historical grounds. For it was widely held in the past, by Linnaeus among others, that each species represents a separate act of the Creator, while the title of Darwin's most famous book *The Origin of Species* has drawn widespread attention to the subject.

A species cannot be adequately defined in simple terms but for most purposes it may be regarded as a group of organisms fertile with one another and capable of producing fully fertile offspring. The latter qualification is essential: the horse and the ass can be crossed with ease but, though the mules so produced are notable for their hardiness, they are sterile.

As Ernst Mayr has stressed in recent years, one species cannot split into two without some degree of geographical isolation. If this be present, selection will gradually adapt the individuals in each area along distinct lines, so accumulating genetic differences between them until they become first partly, and later completely, sterile when crossed.

Several distinct processes contribute to that result but only the most important of them need be mentioned here. As explained in Chapter XIV (Genetics), the manufacture of the reproductive cells includes a stage in which each chromosome pairs with its corresponding (homologous) member, owing to an attraction between the similar genes which they carry. But if the two parents, which respectively contribute the homologous chromosome-sets, are derived from populations which have evolved independently, there may not be sufficient similarity between their genes to ensure the attraction necessary for chromosome-pairing to take place. The mechanism then fails, and the individual is sterile because it cannot form its reproductive cells properly.

Once this state has been reached speciation will continue even if the geographical barriers which produce it break down, for sterility isolates individuals more effectively than any boundaries of habitat.

The statement that geographical isolation is necessary for species formation refers to the sub-division of one species into two. On the other hand, one species can gradually change into another during geological time, and there is an exceptional type of evolution which allows the production of a new species by hybridization though, for technical reasons, it is almost wholly restricted to plants.

Accelerated Evolution

Certain events are capable of accelerating the evolutionary

processes very considerably. The great fluctuations in numbers to which many populations are subject certainly do so. At a time when more individuals survive than is normal, opportunities exist for the spread of genes which would otherwise be eliminated, so allowing new genetic combinations to take place some of which may be of advantage. They will therefore be preserved when stricter elimination sets in.

From 1913 to 1935 I studied an isolated colony of the Marsh Fritillary Butterfly, *Euphydryas aurinia*, which underwent such a fluctuation in numbers. The species was rare and constant in appearance until 1920, but during the following four years the population increased very rapidly. At the same time it became extremely variable, hardly two specimens being alike. In 1925 the rapid numerical increase ceased and so did the great outburst of variability. The butterfly subsequently remained abundant but settled down once more to a relatively constant form which, however, differed from that which existed before 1920. An opportunity had been given for rapid evolutionary adjustment, and clearly that opportunity had been taken.

Another situation favouring rapid evolution is the sub-division of a population into a number of isolated or semi-isolated groups. It does so not as some, following Sewall Wright, have thought because this allows non-adaptive changes to occur, but because it provides opportunities for selection to adjust a species to the diverse requirements of varied habitats.

These are all minor types of evolution. Those of a more fundamental nature are necessarily slow, though doubtless always in progress. Yet, as pointed out by Julian Huxley, it does seem possible that the greatest evolutionary changes, responsible for establishing the principal groups of plants and animals, have ceased. It may be that geological events have favoured them at certain epochs, and that they may do so again. But it certainly seems true that an immense time has elapsed since any of the chief subdivisions of living organisms arose.

Natural Selection and its Consequences

Darwin himself appreciated, in a way that his immediate followers did not, that evolution is usually a failure. The process that constantly selects the best adapted individuals and spreads their qualities through the population leads almost inevitably to destruction. For this there are several reasons, but only one can be mentioned here. Selection ensures that those organisms best suited to their environment shall contribute most to posterity; they thus become more and more perfectly adjusted to it and less able to live in any other. But environments are not everlasting; sooner or later they will change, and the types that have been most accurately fitted to live in them will be too specialized to follow new trends and will become extinct. Those forms that have retained more generalized characteristics are then adjusted by selection to fill the gaps so caused, with the same final consequences. It is certain that this cycle of events is constantly repeated in the course of evolution.

The form of inheritance assumed by Darwin was one in which the hereditary units blend when brought together, instead of maintaining their identity as we know they do. He realized that this would of itself produce a rapid decline in variability which, since no such decline is apparent, must be balanced by 'new variation' or, as we now say, by a very high mutation rate. Consequently he saw nothing incongruous, as we do, in those theories of evolution which are supposed to work by modifying the structure of the hereditary material; that is, by controlling mutation.

Of such a kind are those of Buffon, who held that the effects produced upon the body by any type of environmental change are inherited, and of Lamarck who founded his views upon the inheritance of the effects of use and disuse. To be effective as evolutionary methods these theories not only require that mutation should be extremely frequent, a conclusion at complete variance with the facts, but that it should be *directed*. It does not suffice that environmental changes should merely *induce* mutation, they must also cause the hereditary material so to mutate as to reproduce in subsequent generations their own effects upon the body. X-rays may modify the growth of hair and they certainly produce mutations, but not necessarily affecting hair growth: there is no means of ensuring that one rather than another gene shall be hit by an a-particle, or modified by any other external agency.

There have in the past been many supporters of Lamarckian and similar forms of evolution, which they accepted because they entertained misconceptions in regard to the relevant facts. Today those heresies have attained vast proportions and publicity for very different reasons: they are sponsored by Communist Russia in opposition to 'Western' science and for political ends. Elsewhere the discoveries of science are welcomed as revealing truth wherever they may lead. The labours of a century have not only supplied the proof of evolution but demonstrated the mechanism responsible for it, and the conclusion that it is controlled not on Lamarckian lines but by selection operating on 'particulate' inheritance is now an established fact.

CHAPTER XII

THE COMING OF MAN

by F. E. Zeuner, D.SC., PH.D.

THE middle of the nineteenth century was a period of re-orientation of views concerning the evolution of Man. It is often assumed that the discovery, so to speak, of man's descent from 'the apes' was due to Darwin, and in particular to the publication of his *Origin of Species* in 1859. In this book, however, he merely mentioned the subject in the most casual manner, and it was not until 1871 that he put his views on human origins before the public. With his usual modesty, Darwin acknowledged his indebtedness to predecessors in the field, and in particular to Haeckel who, as professor of zoology in Jena, had undertaken to expound Darwin's ideas of evolution. Haeckel had been keen on the origin of man. In his *Natural History of Creation* he had developed a system which satisfied the requirements of the natural sciences of the 1860s, and when in 1871 Darwin added his *Descent of Man* to the literature of the subject he did so in order to discuss a particular aspect of the mechanism of evolution, namely, sexual selection.

The importance which man's suggested descent from apes assumed in the minds of educated people in the early part of the second half of the last century was almost solely the result of the two books written by Darwin and by Haeckel. But the conception of a natural evolution of man from some ape-like mammalian ancestor was then already well over seventy years old. After some tentative remarks by the German poet and natural philosopher, Goethe, and by the French scientist, Buffon, it was put forward clearly for the first time by the French biologist, Lamarck, in 1801 and accepted by his contemporary, E. Geoffroy de St. Hilaire (the elder). We are not here concerned with the causes of evolution which these workers believed to be at work, but with the principle of human evolution. The time was not yet ripe for the acceptance

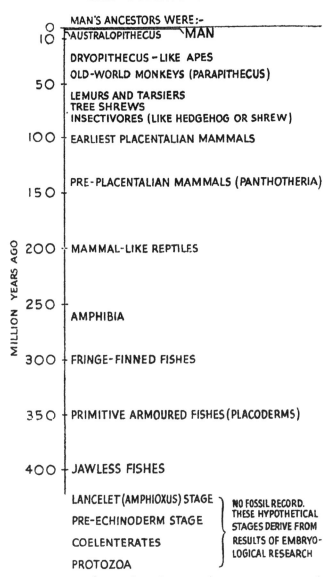

MAN'S ANCESTORS WERE:-

MILLION YEARS AGO	
O	
10	AUSTRALOPITHECUS ⟍MAN
	DRYOPITHECUS – LIKE APES
	OLD-WORLD MONKEYS (PARAPITHECUS)
50	
	LEMURS AND TARSIERS
	TREE SHREWS
	INSECTIVORES (LIKE HEDGEHOG OR SHREW)
100	EARLIEST PLACENTALIAN MAMMALS
	PRE-PLACENTALIAN MAMMALS (PANTHOTHERIA)
150	
200	MAMMAL-LIKE REPTILES
250	
	AMPHIBIA
300	FRINGE-FINNED FISHES
350	PRIMITIVE ARMOURED FISHES (PLACODERMS)
400	JAWLESS FISHES
	LANCELET (AMPHIOXUS) STAGE
	PRE-ECHINODERM STAGE
	COELENTERATES
	PROTOZOA

NO FOSSIL RECORD. THESE HYPOTHETICAL STAGES DERIVE FROM RESULTS OF EMBRYO-LOGICAL RESEARCH

Summary of animal evolution culminating in man; the extremely short period of man's existence in the whole scale of time since the beginning of life is notable.

of such an unconventional idea. This was so, not merely because it was regarded as conflicting with the literal interpretation of the Bible story, but because the great French palaeontologist, Cuvier, dominated the natural sciences with his hypothesis of repeated revolutions and recreations. After the great debate between Cuvier and Geoffroy de St. Hilaire on 19th July, 1830, in which the former's superior knowledge of fossils caused a great impression, belief in

Cuvier's authority became so great that divergent views were relegated to the background.

The Bridgewater Treatises

In the following years, a remarkable series of publications appeared in England. Written by foremost authorities in their subjects, the *Bridgewater Treatises* were intended to demonstrate 'the Power, Wisdom and Goodness of God as manifested in the Creation'. Very naturally, the majority of authors were strongly influenced both by Cuvier's hypothesis of revolutions and phases of new creation, and by the more or less literal interpretation of the Creation stories contained in the *Book of Genesis*. It must be stressed that the defenders of the theory of the special creation of man seemed to have an impressive piece of evidence in their hands. It was that truly fossil bones of man had never been found, either alone or in association with those of extinct animals. At least this was the considered opinion of geologists.

Dean Buckland, professor of geology in Oxford and author of the volume on his subject in the *Bridgewater Treatises*, examined the reputed finds of supposedly early remains of man and concluded that there were not any to be regarded as trustworthy. And as a cautious man he discarded unsatisfactory evidence. So, together with 'Guadaloupe Man' (burials of very recent Indian origin), the reputed finds of fossil man from caves in France were rejected, because 'many of these caverns have been inhabited by savage tribes who, for convenience of occupation, have repeatedly disturbed portions of soil in which their predecessors may have been buried'. The modern archaeologist is only too familiar with this factor of disturbance of natural stratification, and will hardly blame Buckland for his reticence.

Another great English geologist, Lyell, was at that time in full agreement with Buckland. Both, in fact, rejected the hypothesis of *Transmutations* (i.e. the theory of evolution) as applied not only to man but to other forms of life also. Curiously enough, Lyell, though he agreed at that time with Cuvier's views on the creation of man, was his chief and successful opponent in the field of geology. It was he who taught gradual changes in place of sudden catastrophes and who realized that, given enough time, the normal forces of

Nature would bring about gradually those changes which Cuvier had dramatized into awe-inspiring floods, volcanic eruptions and other 'revolutions'.

New Ideas on Geological Time

At this point, the time factor became critically important. Both Lyell and Buckland had convinced themselves and many others that very long periods of time were necessary for geological changes to take place. It was generally agreed that the six-day Creation story must not be understood in the strict chronological sense. This was the first breach laid in the wall of the house of the literal interpreters of Scripture. If time might be interpreted vaguely, for instance in the sense of the 90th Psalm ("A thousand years in Thy sight are but as yesterday when it is past. . . ."), why not the creation of man himself?

The wider and grander views offered by this new kind of natural theology were, however, appreciated by few. One of the few was Charles Babbage, the writer of what he called the *Ninth Bridgewater Treatise*, an independent work. In this he showed how literal interpretation was apt to destroy not only the depth of the spiritual message conveyed by re-translated and oft-copied words, but would damage the reputation of religion as a whole.

"What, then", he asked, "have those accomplished who have restricted the Mosaic account of Creation to that diminutive period and who have imprudently rejected the evidence of the senses? They have succeeded in convincing themselves or others, that one side of the question must be given up as untenable. Those who are so convinced are bound to reject that which rests on testimony, not that which is supported by still existing facts."

Thus, by the end of the first half of the last century, some ground had been prepared for the acceptance of the theory of the evolution of man.

Proving Human Evolution

Having passed the half-century mark, we find a changing situation. The work of Wallace, Darwin and Haeckel was demonstrating the possibility, nay, the likelihood, of human evolution. But still the evidence from human fossils was not regarded as convincing.

What, then, was the kind of fossil evidence for early man that was known about the middle of the last century? The first human skeleton placed before the scientists as of Pleistocene age was found in 1823 by Ami Boué in the loess of the Rhine valley. It became the victim of Cuvier's criticism. Many other finds were made subsequently, but were all discarded as burials or intrusions. Since they were all recovered in a more or less unscientific manner, it was only natural that not one of them could be accepted as proving the great antiquity of man beyond dispute. Moreover, all these fossil men were so like modern man that, if anything, they suggested that no evolution had taken place in our species since it first appeared on earth.

Neanderthal Man

To establish the fact of human evolution a type of fossil man had to be unearthed which was more primitive in structure than modern man. This great discovery was actually made when the famous Neanderthal man came to light in 1856, not far from Dusseldorf in Rhenish Prussia. The cranium was exceedingly thick, and had a low forehead and protruding brow-ridges. It was indeed that of a type which might have been intermediate between man and ape. It was hailed, therefore, by the growing number of evolutionists of the day, but it must be noted that Darwin had not yet published *The Origin of Species*. Critics were quick to appear on the scene: the most formidable was the great anatomist Virchow, of Berlin, who declared that this cranium was clearly pathological and that its owner had been an idiot. But this was the last occasion when the anti-evolutionists thought that they had pricked another bubble and that it all was a mighty joke.

The reason for the general change of attitude was not, however, that further and more convincing discoveries of fossil man tipped the balance. Nor, curiously enough, was human evolution demonstrated with the aid of palaeontological evidence by the evolutionists themselves. Darwin did not make use of Neanderthal man, perhaps because he was too cautious, nor did Haeckel, who was more willing to draw conclusions somewhat light-heartedly, base his claims on the fossils. No, Haeckel claimed to have proved human evolution from the ontogenetic stages of the human embryo. And in palaeontology, the tide was turned by the finds of Stone Age tools of man in the north of France in the forties of the century. We will discuss this fascinating matter in Chapter XIII; here it is necessary to provide a picture of the further development of our knowledge of the physical evolution of man.

The discovery of Neanderthal man was supplemented by the chinless jaw from La Naulette in Belgium. It was found in 1865. In 1887, other Belgian caves, at Spy near Namur, furnished two skeletons which finally disposed of the suggestion that the characters of the Neanderthal race were due to pathological causes. Thereafter, new finds were made every few years, the most important being those of La Chapelle-aux-Saints in France (studied by Marcellin Boule), Krapina in Croatia, Steinheim and Ehringsdorf in Germany, Saccopastore and Monte Circeo in Italy. Many other Neanderthal remains have been found, some as far apart as Palestine (Galilee and Mount Carmel), Siberia and Gibraltar. As the material increased, comparative studies became possible, such as the very recent work of S. Sergi in Rome on the Neanderthal skulls from Italy.

Java Man or Pithecanthropus

In 1891 a new discovery was made which appeared to confirm the by now popular theory of the descent of man. It was *Pithecan-thropus* from Trinil in Java. The cranium, with its very heavy brow-ridges above which there was no forehead at all, was so strikingly intermediate between man and ape and moreover more primitive than in Neanderthal man that it fitted beautifully into the phylogenetic tree. It became now apparent that the lineage of man may have run from apes through *Pithecanthropus*, followed by Neanderthal man, to *Homo sapiens*.

But with the cranium was found a femur. This caused considerable trouble. The femur was evidently much like that of a *Homo sapiens*. In fact it looked more modern than that of Neanderthal man. So the question was raised whether this femur was after all not that of *Pithecanthropus* but a later 'intrusion'. Moreover, as it has an abnormal outgrowth on the upper portion of the shaft, pathology was once more invoked to belittle its importance. Today, it is agreed that the growth on the femur of *Pithecanthropus*, though pathological, is not likely to have influenced its shape significantly. This view has been greatly strengthened by the more recent finds.

Pekin Man or Sinanthropus

These finds consist of two groups. The members of the first come from several localities within a radius of about forty miles from Trinil where they were collected by the Dutch geologist, von Koenigswald. There are a massive lower jaw, a cranium more complete than the original one, several other specimens and, last but not least, a baby skull of a *Pithecanthropus*. This came from Modjokerto. The second group was found in China, in the neighbourhood of Pekin. These finds, which were called *Sinanthropus*, were made by the British anatomist, Davidson Black, and the Chinese palaeontologist, Pei, and comprise fragments from as many as forty individuals. The Chinese and Javanese groups are closely related and represent no more than races of one type of primitive man.

The mammalian fauna found in association with the human bones, both in Java and in China, was sufficient to determine their geological age. It is approximately Lower Pleistocene. The *Pithecanthropus* group was thus much older than any Neanderthal find. Though most of the supplementary discoveries date only from the last two decades, the chronological order of *Pithecanthropus* first, Neanderthal man second, *Homo sapiens* third, had been established by the turn of the century. It tallied with most of the anatomical evidence, derived from the fossils, for a progressive evolution of man. It was a highly satisfactory state of affairs.

More skulls from Europe

As so often in science, the simple solution turns out to be the wrong one, and it is a very human trait that some of the workers hesitate, or even refuse, to accept a modification of their original ideas suggested by more recent discoveries. In the case of fossil evidence for human evolution, difficulties began to accumulate from the year 1912 onwards when the Piltdown skull and jaw were found in Sussex. As is well known, the skull resembles *Homo sapiens* in many respects; in particular, there are no brow-ridges. But the jaw is amazingly ape-like. Many found it hard to believe that these two pieces could come from the same individual. The controversy which began over Piltdown man has continued to the present day, and its ultimate outcome is still uncertain.

In 1912 and the years immediately following, however, this specimen caused some uneasiness. The accompanying fauna appeared to suggest a Lower Pleistocene age (this is no longer believed to be so, and an Upper Pleistocene age is now favoured), but the sapientoid skull did not fit the conceptions of some experts of the time, who would have preferred to find the characters of Pithecanthropus in this fossil. On the other hand, if one assumed an Upper Pleistocene age, the exceedingly ape-like jaw proved to be the stumbling block. The controversial nature of this specimen, however, saved the human lineage from a surgical operation for awhile.

In 1933 an almost complete skull was found in the mammal-bearing deposits at Steinheim an der Murr, not far from Stuttgart. It appears to have belonged to a young female. Its brow-ridges are very heavy, and there are other neanderthaloid characters. But the face is not prognathous, the back of the head is rounded, and the wisdom-tooth is noticeably reduced. In these and other respects the Steinheim skull was suggestive of *Homo sapiens*. The presence of the latter group of characters was strange indeed, since the specimen dated from the Last Interglacial and was therefore older than the majority of the Neanderthalian specimens. Here then was a fossil that did not fit into the prescribed scheme of human evolution, though admittedly it was not very far out.

But only two years later, another specimen came to light which, although very incomplete, showed *sapiens*-like characters but was considerably older than the Steinheim skull. This was the fragmentary Swanscombe skull, found by A. T. Marston in a gravel-pit near Gravesend, Kent. The age of the gravels containing this fossil is the Penultimate Interglacial, which is separated from the period of Neanderthal man by an entire glaciation. Osteologically, the fragments proved to be so much like modern man that it would not have been justifiable to give them a special scientific name. It was pointed out, however, that they might have come from a specimen of the Steinheim type and therefore have been combined with certain neanderthaloid features. Yet, there was by now not the slightest doubt that certain types of man which were older than the typical Neanderthaler were more *sapiens*-like than the latter.

The Finds from Mount Carmel

Further confirmation of this new observation came from Palestine, where about a dozen fossil men had been found in the early thirties. The majority were excavated by Dorothy Garrod in caves of Mount Carmel, and studied by Keith and McCown. There was a whole range of types, from typical Neanderthalers to others approaching *Homo sapiens*. Two interpretations have been put forward. In their earlier publications, Keith and McCown assumed two races (Neanderthaler and *Homo sapiens*) which were intermingling. Later on, however, they submitted that the Palestine group was a genuine, transitional race linking Neanderthal man with *Homo sapiens*. At the time it seemed too daring to assume the existence of *sapiens*-like man at a period when he could mix with Neanderthal man. (It must be remembered that the discoveries of Mount Carmel, Steinheim and Swanscombe were almost simultaneous.)

Today, after a lapse of fifteen years, the hypothesis of interbreeding is gaining the ascendency. The Mount Carmel group is racially so heterogeneous that it is difficult to regard it as pure. Its age is Last Interglacial: it is contemporary with, or later than, the Steinheim skull with its *sapiens*-characters, and the associated Stone Age industries also render the view of a mixture of two races tenable.

Neanderthal Sub-types

Meanwhile the large number of Neanderthal specimens which had accumulated in the collections was calling for a comprehensive treatment. The work of Boule and Vallois in France and, quite recently, that of Sergi who started from the Italian finds of Neanderthal man, have shown quite clearly that the early Neanderthaler (who lived in the Last Interglacial under temperate conditions) was in several respects less extreme in his characters and therefore a little more like *Homo sapiens* than the late Neanderthaler. The finds at Krapina in Croatia, Taubach and Ehringsdorf near Weimar, and Saccopastore near Rome, belong to this early group. The late group are the 'typical' Neanderthalers of La Chapelle-aux-Saints, Monte Circeo, etc., all belonging to the first phase of the Last Glaciation.

Can the Neanderthal type of man be traced further back into the past? On this point, evidence is still scanty and unsatisfactory. There is the Steinheim skull which is, perhaps, older than the Last Interglacial. And there is the Mauer jaw—the so-called Heidelberg man—which dates from the Antepenultimate Glaciation or thereabouts and is therefore older than any of the remains thus far discussed. In spite of its massiveness and great size, this jaw would go well with any of the Neanderthal skulls. But the shape of the jaw does not enable us to make pronouncements on the skull to which it belonged, so that the matter remains in suspense. Some physical anthropologists, like Weidenreich, however, are inclined to assign the Mauer jaw to the Neanderthal group. Should this prove to be correct, Neanderthal man may have to be derived from the *Pithecanthropus* type of the Lower Pleistocene.

The Coming of "Homo sapiens"

The great problem that remains to be discussed is the origin of *Homo sapiens* himself. The fossils have shown that he is not descended from the typical Neanderthal man. Moreover, the investigation of prehistoric sites has made it clear that in Europe Neanderthal man was followed rather abruptly by *Homo sapiens*, who was the bearer of a new culture called the Upper Palaeolithic. Everything points to his having invaded Europe and superseded Neanderthal man in a comparatively short period of time. Coon holds the view that in this process a certain amount of Neanderthal blood was absorbed since the earliest Upper Palaeolithic men, as represented by the find of Combe Capelle in France, were in some respects more extreme '*sapiens*' than their descendants, such as Cro-Magnon and Predmost man of the Aurignacian, and Laugerie Basse and others during the Magdalenian.

It has already been pointed out that the roots of *Homo sapiens* may be found in *primitive* Neanderthal types, such as Steinheim man. But the lineage might easily go much further back into the past. The two skull fragments from Fontéchevade in France, which were discovered by Mademoiselle Henri-Martin in August 1947 and have since been studied by Vallois, have made this very probable. Their geological age appears to be Last Interglacial, so they are approximately contemporary with the early Neanderthal group. But they are very different and approach Swanscombe man which is much older. They resemble modern man more than the Neanderthaler does, but are not identical with him. This fresh evidence renders it highly probable that the lineage of *Homo sapiens* goes back to at least the Middle Pleistocene, without merging with the *Homo neanderthalensis* lineage.

Whether both lineages originated from Pithecanthropus of the Lower Pleistocene, cannot yet be decided. There are indeed claims, which require further substantiation, that *sapiens*-like man is as old as the Lower Pleistocene. Leakey, for instance, believes that a jaw of this type from Kanam near Lake Victoria is of Lower Pleistocene age.

Some Deviations

Human evolution, as we see it today, thus provides a picture rather different from the original conception. Regarding the lineage of *Homo sapiens* as the main one, Neanderthal man forms an extinct side-branch.

There may well have been other extinct side-branches. Several fossils have been found which do not fit into either the *sapiens* or the *neanderthalensis* line. A *Pithecanthropus* was found in Java (*P. robustus*) which has a very large and heavy skull, and a gap between the front teeth and eye-teeth which is characteristic of apes and monkeys but is absent from all humans, fossil and recent.

There is other evidence for gigantism in early man. The large size of the jaw of Heidelberg man (from Mauer) has been mentioned. In respect to size, however, the most remarkable specimen that has yet come to light is the fragment of a lower jaw of *Meganthropus*. Von Koenigswald obtained the specimen just before the Japanese occupied Java, and a preliminary description was subsequently published by Weidenreich. For a human jaw it is truly enormous, the teeth themselves participate in this gigantism. It would fit a strong full-grown gorilla. We therefore have to reckon with a far greater variety of extinct humans than has hitherto been considered possible.

Sub-human Fossils

Up to this point we have confined our attention to evidence for the physical evolution of Man in the strict sense. Two further problems remain to be discussed, namely, first, whether there are sub-human types of fossils linking the most primitive types of man known with the ape stock from which he is supposed to have arisen; and secondly, how did the modern races of mankind come into existence?

As regards possible sub-human ancestors of man, it is remarkable that evidence had been missing or misinterpreted until a few years ago. In this respect, the middle of the last century was still a complete blank, although some of the primitive apes of the *Dryopithecus* group, which are now regarded as near the origins of the human stock low down at the level of the Mid Tertiary apes, had actually been described. Of this group, large numbers of specimens have been recovered from beds of Lower Miocene age (about 20–25 million years ago) occurring on Rusinga Island on Lake Victoria. The best-known type is called *Proconsul*. It was first described from a small fragment by Hopwood eighteen years ago, but in 1948 Mrs. Leakey found an almost complete skull. This and the other remains have been studied by Le Gros Clark, according to whom there can be little doubt that in this group are to be found the ancestors of all the modern large apes and of man.

So far (and it must be emphasized that most finds were made quite recently) it appears that the Dryopithecinae originated in Africa. The view may therefore be held, for the time being, that the first steps in the evolution of mankind occured in Africa. But for how long can this be maintained? In all probability only until equally intensive field-work is undertaken in another part of the world. The honour of being called the mother country of mankind was first conferred upon Europe where the first finds were made, and later on south-east Asia, where the *Pithecanthropus* group was unearthed. Now it is Africa's turn. Which country's turn will it be next? Perhaps India's?

Australopithecus—The Missing Link?

It is indeed astonishing how much material Africa has yielded within the last few years, for there is a second group of fossils which at the moment occupies the first place in discussions on human evolution. It comes from South Africa. The specimens have received a variety of names but are most conveniently combined as *Australopithecus*. The first find was made in 1925 by Dart, who recognized in it a remarkable combination of characters of ape and man. Unfortunately, his views were not favourably received by the majority of anthropologists at the time. New finds were required to convince the experts, and they began to pour forth from 1936 onwards. Most of them were made by Dr. Broom in a few localities not far from Johannesburg, and they have been studied since 1947 by Le Gros Clark.

The *Australopithecus* group indeed deserves the title of 'The Missing Link' more than any other fossil. This is so from the osteological point of view. It is therefore most regrettable that it has not yet been possible to ascertain its geological age, though the specimens were associated with extinct mammals. The skull of the Australopithecines quite superficially resembles that of a chimpanzee, with its projecting and heavy jaws. But there are highly significant differences in the brow-ridges which, though heavy, do not form the bony shelf encountered in the apes; in the teeth which are exceedingly human; in the position of the line of attachment of the neck-muscles; and in the position of the occipital condyles (which connect the skull with the vertebral column). The last two characters are very startling, for they suggest that the head was carried after the manner of man. Body posture, therefore, appears to have been upright.

Confirmation is provided by the hip bone, which has a very broad blade as in man and is distinct from the narrow one characteristic of the apes. It is virtually certain, then, that *Australopithecus* stood and moved about much like a man, and it is to be inferred that his arms were no longer used for locomotion. Moreover, this remarkable creature, with the face of a chimpanzee and the body structure of a man, lived on the ground in an open country. In this respect, too, he differed from the modern apes, which are restricted to forests.

Looking Back

Whilst palaeontologists were studying the fossils themselves, geologists investigated their chronological sequence and age. Geochronological research has developed several methods of dating in years. These methods can often be applied to the fossils in a round-about fashion only and they are in themselves not very accurate. Nevertheless, they do give a fairly good idea of the rate at which changes occurred in the succession of descendants leading up to modern man, and it has been thought worth while to include them in the summary contained in the following paragraphs. They provide, in a few words, the state of our knowledge in the year 1951, as contrasted with 1851 and 1901.

By 1851, only the idea of human evolution had been put forward. It had been rejected by the majority of both scientists and laymen. By 1901, the idea that man sprang from the apes had been universally accepted and become popular, so much so that the few surviving species of apes were regarded as closer to our ancestral line than was justified by osteology. Moreover, enough fossil evidence had been discovered to reconstruct a lineage leading from an unknown ape through *Pithecanthropus* and Neanderthal man to *Homo sapiens*.

In 1951, it can no longer be said that this view stands on firm ground. In particular, it now appears necessary to exclude Neander-thal man from our direct lineage. But the field has been enormously widened by the finds of sub-human types of Hominids and apes, and comparative anatomy has added, in a generalized way, the groups representing the stages of our most distant ancestry. Thus it looks as if, about 350 million years ago, our ancestors were jawless fishes. About 250 million years ago, they had become amphibians. Another 50 million years later, they were 'mammal-like reptiles'. About 120 million years ago they had reached the mammal stage and were soon after to develop that characteristic organ of the mammals, the placenta. About 70 million years ago, at the transition from the Cretaceous to the Tertiary period, man's fore-runners were tree-shrews, small animals superficially resembling squirrels, but related rather to insectivores like the hedgehogs. This group, incidentally, is already so much like the Primates, that it has recently been included in them. Another ten million years later, the level of the tarsiers had been reached, and again ten million years on, that of true monkeys. About 20 to 25 million years ago, our ancestors were to be found among the Dryopithecine apes, as represented for instance by the East African *Proconsul* group. Here for the first time some 'human' traits may be discerned; in particular, these apes were of a lighter build than modern ones and better adapted to running and walking on the ground.

By Pliocene times, the *Australopithecus* stage must have been reached, and adaptation to bipedal locomotion was complete. The hands had become free to take over their manifold specifically human functions. This stage cannot, unfortunately, be dated as yet. About half a million years ago, we encounter the Pithecanthropus group, definitely primitive man. But it is not certain whether the lineage of modern man went through *Pithecanthropus*. About a quarter of a million years ago, Swanscombe man shows the first *Homo sapiens* characters. Neanderthal man is in all probability a side-branch which appeared in Europe with an unspecialized type about 150,000 years ago and the specialized type of which became extinct in or after the Last Glaciation, when modern man colonized Europe. This was about 80,000–100,000 years ago.

From the point of view of evolution, the remarkable outcome of a century's work is that the upright posture of man is clearly much older than the great development of his brain. Palaeontological evidence is unambiguous in this respect, and theories giving precedence to the brain, once flourishing and last propounded by Dabelow in 1931, have to be discarded.

'Human palaeontology', the study of fossil remains of man, was non-existent in 1851. A hundred years of fortunate discoveries and their investigation have put this branch of learning, which combines anthropology and archaeology with palaeontology and geology, on its feet. As a borderline subject, it exemplifies a fact which has in the course of a century become apparent in many other fields of research, namely, that the combination of diverse lines of approach often produces more satisfactory results than the orthodox one of confining work to one particular line. Moreover, dealing as it does with the origins of our own species, its philosophical importance is considerable.

GEOLOGICAL HISTORY OF MAN

Tentative time-scale in years	Period	Climatic phase	Man and his ancestors
(*Before Present*) 15,000	Holocene	Postglacial	} }Homo sapiens }
		Last Glaciation	}
100,000	Upper Pleistocene		}Neanderthal Man
150,000		Last Interglacial	} Fontéchevade Man
	Middle Pleistocene	Penultimate Glaciation	
250,000		Great Interglacial	Swanscombe Man
500,000	Lower Pleistocene	Antepenultimate Glaciation Antepenultimate Interglacial Early Glaciation	Heidelberg Man *Pithecanthropus* *Sinanthropus*
600,000			
		Villafranchian	*Australopithecus* (age conjectural) *Dryopithecus*
	Pliocene		
c. 12 million			
c. 25 million	Miocene		*Proconsul, Dryo-pithecus*

CHAPTER XIII

THE PROGRESS OF *HOMO SAPIENS*

by F. E. Zeuner, d.sc., ph.d.

Whilst work on the origins of man has at all times been regarded as a palaeontological problem, the present distribution of *Homo sapiens* has been viewed largely from the biological angle. The conception of *Homo sapiens* as one of the numerous species constituting the present-day fauna of the earth, dates from the early part of the eighteenth century. Even the classification of the existing races of mankind is over two hundred years old. It was developed by Linnaeus who duly incorporated it in his *Systema Naturae*. A hundred years ago, Blumenbach's division into five races was almost universally adopted. This author distinguished the Caucasian, Mongolian, Ethiopian, American and Malayan races. This classification suffered from the neglect of important races in then less known countries, such as the Melanesians and the Australians. The credit of having recognized the basic three-fold division of modern man into the Whites, Yellows and Blacks, is due to Cuvier. The most recent classifications, such as those of Hooton and Montagu, preserve this feature.

But the last hundred years have witnessed a great development in the methods of description of individuals and the generalization of racial characters. The few features used in Blumenbach's day—namely, skin and hair colour, form of hair, shape of the elements of the face, skull proportions, body size, etc.—have been supplemented by numerous exact measurements and descriptive terms. Indices are used to relate proportions as well as combinations of unconnected characters. The system of 'somatotyping' developed by Sheldon, for instance, distinguishes seventy-nine varieties of human physique. In the measuring of physical traits, the skeleton still provides the basis, even when soft parts are investigated. But the description of races is no longer restricted to somatotyping and anthropometry. Physiology is beginning to contribute, and psychology and the study of behaviour are playing their part. Thus, the study of blood-groups has made great strides since 1919, and the so-called Rh (rhesus monkey) - factor has achieved popularity within recent years. The inheritance of these physiological factors is subject to the ordinary rules of genetics, and speculation has been busy trying to interpret this evidence phylogenetically. As to psychology, attempts have been made to classify human temperament in a scientific manner, and experimental work, on both man and apes, has yielded much significant evidence.

The Classification of Races

Nevertheless, the mixing of human races has made it very difficult to obtain a classification based on phylogenetic relationship. Several of the systems in vogue at present still distinguish the three primary races, namely Caucasoid, Mongoloid and Negroid. Ashley Montagu divides the Caucasoids proper which range from the Nordics and Mediterraneans to the Polynesians. The Negroids comprise both the African and Asiatic-Oceanic sub-groups. The importance of crossing of races is generally recognized: the Indo-Dravidians, for instance are interpreted as a mixture of Mediterranean Whites with older Australoid and Negroid elements, and the Bushman-Hottentots, or Khoisanids as von Eickstedt calls them, as the descendants of the ancient South African Boskop type which has absorbed Negrito and even Mongoloid blood.

History of Modern Races

The last-mentioned instance illustrates once more that the evidence needed is palaeontological. Some material is already available. It does not speak plainly and unambiguously, but has stimulated healthy speculation. As to Europe, Coon holds that the first arrivals of *Homo sapiens* in Europe after the reign of the Neanderthaler were true Whites (as witnessed by the fossil from Combe Capelle) and that they subsequently deteriorated somewhat by taking up Neanderthal blood. The White race therefore must have originated elsewhere, presumably in Asia, prior to the second phase of the Last Glaciation. Soon after the appearance of these Whites in Western Europe, racial differentiation became apparent.

The Cro-Magnon type of Upper Palaeolithic man, of which about thirty-five skulls are known, has been studied by Morant. It appears that this group comprised an eastern, long-headed, type and a more broad-headed western one. But in addition, some racial oddments have been found in the Upper Palaeolithic of Europe, such as the so-called negroids of Grimaldi on the Italo-French Riviera which Sollas and Boule tried to link with the Bushman. Today, their negroid characteristics are once more considered unconvincing and *sub judice*, for no others of their type have been found. There is also the Chancelade skull, associated with a Magdalenian culture, and a few others, which exhibit eskimoid features. No agreement has been reached as to their significance. Whilst Coon and Hooton regard the eskimoid features as merely due to convergence, Morant and Sollas think of a real racial relationship. In spite of the relatively abundant fossil material, the origin of the surviving major sub-races of the Whites is still obscure. It appears, however, that they originated north of the great Asiatic mountain belt.

Curious fossil evidence has been found in the Upper Cave of Chou-kou-tien near Pekin. Of the several Upper Palaeolithic individuals buried together one was difficult to define, perhaps a primitive Mongoloid (this is Weidenreich's view), perhaps an archaeic Australoid or even an Ainu (Hooton's view). But it was not a Chinese. Another skull is believed to have Oceanic negroid (Melanesian) relations, and yet another looks like an Eskimo. A strange ensemble of racial elements indeed, but there is at least the suggestion of a primitive mongoloid element, and they do not conflict with the widely held view that the mongoloids became differentiated in East Africa.

From Java, two sets of fossils later than *Pithecanthropus* are now known to correspond approximately to the Neanderthal and *Homo sapiens* stages—Solo man and Wadjak man respectively. The latter is important because of its australoid characters, suggesting that the Australians did come from Asia. Weidenreich, incidentally, thinks that they evolved locally straight from *Pithecanthropus*, but few of his colleagues have followed him.

In South Africa, several skulls have been found which again are somewhat suggestive of the australoid. At least one of them, however, the Boskop skull, is reminiscent of the Bushman in having a large brain case and a small face and this in spite of its enormous size. The Boskopoids are therefore regarded as ancestral to the Bushmen-Hottentots, the reduction in body-size which accompanied this evolution being possibly due to the admixture of pigmy blood. It is perhaps significant that the South African fossils make it possible to derive the modern races from them, and also that there seems to be a primitive, australoid element about.

It is strange, however, that no indisputable ancestor of the negro has yet been found. The only specimen which might be regarded as such is the undated yet well fossilized Asselar Man, from the middle of the Sahara. Boule and Vallois regard him as an unspecialized negro ancestral to the forest negroes, but Hooton compares him with the long-limbed Nilotic African.

Comparison with 1851

Where, then, do we stand today as compared with a hundred years ago? In 1850, the *Christian Examiner* published an article by Agassiz in which it was stated that the various races of man were distributed over the globe like distinct species of mammals. For the first time, the principles of zoogeography were here applied to man. In the years that followed, Darwin was at pains to show that the races of man were hardly more than subspecies from the taxonomist's point of view, and this interpretation is shared today both by anthropologists conversant with zoological nomenclature and geneticists like Dobzhansky. This view implies the existence of some ancestral *Homo sapiens* with a number of australoid characters, which spread over the continents and became differentiated into the main or primary races.

For this process, a measure of geographical isolation would appear to have been necessary, and the glaciated areas of the Last Glaciation in Asia are often held responsible. They may indeed have separated the Whites from the Yellows and both from the inhabitants of southern India, perhaps the Australians. The Negroes would have been isolated beyond the dry belt in Africa. Fossil evidence, however, now renders it difficult to maintain such a simple view. *Homo sapiens* appears to be older than the Last Glaciation, and racial differentiation to have started early. Moreover, mixing of races at an early date is suggested. And the geographical picture is spoilt by the existence of negroids both in Africa and Melanesia, so that the assumption of large-scale displacements of entire races has become almost inevitable.

In his *Descent of Man* Darwin eagerly defended the origin of all the races by differentiation from a single ancestral form. He was fighting the hypothesis of Vogt that several ancestral forms were involved in man and that similarities were partly the result of convergence. Today, Darwin's view is still held by the majority of workers and it appears the more reasonable. But the alternative view has not become extinct, since Weidenreich has in recent years upheld a polyphyletic ascent of *Homo sapiens* from different ancestors of the *Pithecanthropus* level. It is evident therefore that, though our knowledge of facts has enormously increased in the last hundred years, our understanding of the process of formation of human races has not increased correspondingly.

Evolution of Technology

In conclusion, it is necessary to refer briefly to another aspect of the study of man, namely, the evolution of technology. Prehistoric research has played a very important part in the forming of our views on human evolution, for it was familiarity with Stone Age tools which a hundred years ago swayed opinion in favour of the antiquity of man. It was Boucher de Perthes who first put forward convincing evidence of the Pleistocene age of certain types of Stone Age tools. These, now classified as Lower Palaeolithic, came from the gravel deposits of the river Somme in North France. Boucher de Perthes's paper of 1846 was followed by his *Antiquités celtiques et antédiluviennes* a year later. But for seven years the tide of opinion was against him. In 1853 he was joined by Rigollot. That prophets are without honour in their own country, applies in this case as in so many others. It was the succession of visits of a number of British geologists and archaeologists which changed general opinion.

The first was that of the palaeontologist Falconer, who saw Boucher's collections at Abbeville. He informed the geologist Prestwich and the archaeologist Evans, who were convinced by the evidence. All three reported to the Royal Society in 1860. Further visits followed, and it was recalled that the British Museum owned a stone tool resembling the French ones but found with an elephant's skeleton in London in 1715, and also that Frere had found others at Hoxne in Suffolk and reported them in *Archaeologia*. By 1865, when Sir John Lubbock published the first edition of his book, *Prehistoric Times*, the antiquity of Stone Age man of the river gravels had been widely accepted, together with that of the Upper Palaeolithic finds made so abundantly in French and Belgian caves. The influence of this information on the theory of evolution was considerable, since it paved the way for the acceptance of the evolution of man in spite of the absence of undisputed fossils.

Since those days the classification of the material cultures of early man has developed a great deal. A chronological sequence has emerged which, in a few words, suggests the following course of events. What might be called the first stage, that of using unfashioned natural objects as tools, goes far back into the anthropoid stage of human evolution. It must have lasted millions of years. One cannot say when the second stage, that of deliberate smashing of stones and of selection of suitable pieces of wood for use as tools, began, but it was certainly back in the Pliocene. *Australopithecus* appears to have been at the first, if not the second, stage. At the beginning of the Pleistocene, about 600,000 years ago, man had entered the third stage of technological evolution: he had learned to flake stone so as to obtain definite shapes. About 400,000 years ago, the technique of making tools had much improved. Another 100,000 years later, further development in the mental abilities of the makers is suggested by the appearance of the complicated Levalloisian technique of flaking stone. But all the cultures of the Lower Palaeolithic were extremely long-lived; the Acheulian, for instance, appears to have lasted for something like 300,000 years.

Towards the end of the Lower, and during the Upper, Palaeolithic, the implements become more variable and display versatility and technological advance. It was the time when *Homo sapiens* occupied Europe. Within that period, cultures had life-times of a few ten-thousand years. To this group may be added the fifteen-odd thousand years of the Mesolithic, which ended about six or seven thousand years ago. Thereafter, acceleration becomes conspicuous— two thousand years of New Stone Age or Neolithic, twelve hundred years of Bronze Age, and finally the Iron Age beginning about

800 B.C. and still continuing, with its quickening pace of technical progress. It appears that the accelerated pace of the technological evolution of man is not matched by a corresponding evolution of his body and brain, at any rate not since the beginning of the Upper Palaeolithic. Since then man has busied himself to learn, and practising of his mental powers has meant increasing intercourse with his fellows and a quickening pace of inventiveness.

Apart from a chronological sequence, prehistory has further revealed the contemporaneity of different cultures in different regions. This fact is rendering the sequence increasingly complex, especially from the Upper Palaeolithic onwards. It is evidently the expression of the presence of distinct ethnic and ecological units in the population. But the linking of cultures with skeletal remains is a difficult matter. This is not surprising, since burials disturb the stratification, and it requires both favourable conditions and careful excavators to identify the implement level to which a skeleton belongs. Some progress has been made in the sorting out of human remains belonging to the late phases of prehistory. This work, which is of very recent date, has stimulated Coon to synthesize our knowledge so far as Europe is concerned. It is evident that close co-ordination of human palaeontology and prehistory is becoming an urgent task of scientific research on man.

If one surveys the field covered by this article, one must agree with Hooton, who I believe once said that new data do not always bring better understanding and that processes which seem explicable to us in our ignorance have with increase in our factual knowledge become complex and mysterious. It is this mystery which is still attached to research bearing on both early and primitive man that has drawn so many workers into the subject.

OUTLINE OF THE PROGRESS OF MANKIND

Approximate date

A.D. 1900	Beginnings of the scientific age.
1800	"Industrial Revolution": machines replace the work of man's hands on a large scale.
Early 16th century	European Iron Age conquering American Stone Age.
900	Viking Period: long-distance sea-travel extends to the Atlantic.
B.C. 4	Birth of Christ.
350	Climax of Greek philosophy (Plato and Aristotle).
6th century	The great Eastern philosophers and religious founders (Jeremiah, Zoroaster, Mahavira, Buddha, Confucius). Beginnings of Greek philosophy (Thales).
650	Iron Age starting in extra-Mediterranean Europe; a cheap metal becomes available.
1500	Moses: the foundation of Monotheism.
1600	Beginnings of the alphabet.
1800	Bronze Age beginning in extra-Mediterranean Europe: metal replaces stone.
2500	Royal cemeteries of Ur.
3000	Establishment of writing. Food production methods spreading to Northern Europe with the Neolithic culture. Menes unites Lower and Upper Egypt.
4000	Invention of the wheel.
5000	Earliest known agricultural civilization in Mesopotamia and adjacent countries.
7000	Hypothetical beginning of food production in Western Asia.
9000	Mesolithic period of the Stone Age beginning in Northern Europe (it was already old elsewhere). Man still a hunter and gatherer of food.
20,000	Last climax of the Last Glaciation, European man in the Upper Palaeolithic stage.

70,000–150,000	Mousterian culture developed by Neanderthal man. He already practised burial rites, which implies belief in spirits.
100,000	*Homo sapiens* developing the Upper Palaeolithic stone 'blade' cultures and subsequently spreading to Europe.
250,000	'Levallois' technique of stone-flaking discovered.
100,000–500,000	Hand-axes the most characteristic tools of Man (Abbevillian or Chellian, and Acheulian cultures).
600,000	First or Günz Glaciation.
700,000	Man uses pebbles and stone chips, but does not yet shape his tools.

CHAPTER XIV

GENETICS AND EMBRYOLOGY

by G. R. DE BEER, M.A., D.SC., F.R.S.

IN the *Origin of Species*, Darwin had found it necessary to say that "the laws governing inheritance are for the most part unknown. No one can say why the same peculiarity in different individuals of the same species, or in different species, is sometimes inherited and sometimes not so; why the child often reverts in certain characters to its grandfather or grandmother or more remote ancestor; . . ." This was published in 1859, and, had Darwin but known it, the answers to his questions were at that time being worked out by Gregor Mendel as a result of his experiments on peas. Mendel's work was published in 1865 but attracted no attention until 1900, by which date comparable results had been independently obtained by other workers, notably Correns, Tchermak, and de Vries. Very shortly afterwards, W. Bateson and R. C. Punnett showed that these results applied to animals as well as to plants.

The gene: the physical basis of heredity

The essence of Mendel's results was the demonstration that the characters which he studied are controlled by inherited factors which in any individual exist in pairs, one having been derived from each parent; that these factors may exist in two *states*, each controlling one type of character, so that of a pair of factors in an individual, both may be of one type or of the other type, or one may be of one type and the other of the other; and that the factors separate from one another (*segregate*) at the formation of the reproductive cells, so that not more than one factor of any pair goes into the same reproductive cell.

Organisms in which both factors of a pair are identical breed true for the character which they control; when the two factors of a pair are different the organism is hybrid for that character and does

O

not breed true. And as organisms display large numbers of characters, Mendel also investigated the distribution of factors between parent and offspring when the parents differed in more than one pair of characters. The result was the demonstration that each pair of factors 'segregates' independently of the others.

The most important consequence of these results is the demonstration that the physical basis of heredity is particulate—a conclusion equivalent in importance to the discovery of the atom in chemistry, the electron and the quantum in physics, and the cell in biology. The Mendelian factors are discrete particles.

The effect of a factor when present in an individual together with another factor of a different type may be either to dominate the other, in which case the factors are known as *dominant* and *recessive* respectively, or to produce an intermediate condition in the character which it controls. But the fundamental point is that whatever the appearance of an organism which has been produced by parents differing in a character, its offspring (the second filial generation, or F_2) will show, by the clean-cut ratio of categories of individuals which it contains, that the factors controlling the character have not 'contaminated' each other but are recoverable, in their complete and original purity, in the grandchildren of the original parents. Here is the answer to Darwin's question.

Mendel's principle of analysis, which is simply to breed together individuals differing in a small number of well-defined characters, has been applied to every group of both the vegetable and animal kingdom, and found to be of universal application. It is not claimed that *all* inheritance (i.e. resemblance between parent and offspring) is Mendelian; but so much is now definitely known to be Mendelian that it is possible and necessary to amend Darwin's words and write "the laws governing inheritance are for the most part known". Very soon after the rediscovery of Mendel's results in 1902, it was pointed out by W. S. Sutton that the behaviour at the formation of the reproductive cells of those constituents of the nuclei of the cell known as *chromosomes*, provides exactly and precisely the mechanical basis required for the distribution of the factors postulated by Mendel's interpretation of his experiments. At the hands of T. H. Morgan and his associates A. H. Sturtevant, C. W. Bridges and H. J. Muller, working on the breeding experiments, and of

C. Stern, T. Dobzhansky and C. D. Darlington, working on the structure and behaviour of chromosomes, the truth of the idea that Mendelian factors are carried in, and undergo the same distributive fate as, the chromosomes has been demonstrated objectively and experimentally with such a wealth of evidence that the theory of the *gene* (as the Mendelian factor is now called) may be regarded as one of the most firmly established pillars of modern biology.

When the early experiments of Mendelian genetics were first performed, it appeared as if the effects of differences in the genes were always reflected in marked and discontinuous differences in the characters which they controlled. With the refinement of analysis applied by R. A. Fisher, E. B. Ford and S. C. Harland, among others, it is now clear that the effects of any gene in controlling a character of the organism are themselves under the control of the other genes present in the organism, the so-called *gene-complex*. These other genes can modify the effect of a given gene so as to render its manifestation more or less marked, with the result that differences in the gene may be attended by small and gradual differences in the resulting character of the organism. Changes in the gene-complex are therefore capable of changing the manifestation of a given gene, and this principle is of great importance since it is believed to be the manner in which evolutionary changes take place.

Modifications and Mutations

A great step forward was effected in the understanding of the principles of genetics by the experiments performed by Johannsen in inbreeding beans for many generations and obtaining plants which breed 'true' for all their genes. A stock of such organisms, whether plants or animals, is known as a *pure line*. By testing individuals of a pure line under different environmental conditions it is possible to discern which changes in their characters are due to the effects of the environment (changes known as *modifications*, which sub-sequent generations do not show unless they are subjected to environmental conditions similar to those which originally resulted in these modifications) and which changes in their characters are due to events which have taken place within the organism (changes known as *mutations*, which subsequent generations do show in environmental conditions similar to those which prevailed before the change occurred). In simpler words, variation in character may be due in the lifetime of an individual to the action of the environment, when such variation is known as a modification and is not inherited; or it may be due to an event internal to the organism, when it is known as a mutation and is inherited.

A mutation in its simplest form is a change in a gene which then continues constant in its changed form until it mutates again. A mutation is therefore the introduction of a novelty in the genetic constitution of a lineage. Something is known of the frequency with which genes undergo mutation; it may reach a level of one in half a million, and this rate may be accelerated by the action of X-rays and radioactive substances. But nothing is known of the conditions in which a mutation occurs or of the relation between such conditions and the effects which the mutant gene shows in the resultant organism. If, as there is reason to suppose, genes are complex protein molecules, they present two puzzling phenomena which have so far baffled analysis; the first is why they are so stable, i.e. why they mutate so infrequently, and secondly why when mutation does occur it is found to affect only one of the two corresponding genes in an organism, although they must be situated very close together indeed in the nucleus of the reproductive cells.

The Determination of sex

A special aspect of the problems of genetics is that of the mechanism whereby it is determined that of the offspring of two parents some will be male and some female. The experiments of C. W. Bridges on the fly *Drosophila*, involving both breeding and investigation of the chromosomes, have demonstrated conclusively that the determination of sex in that form is the result of a balance between the genes carried in a particular pair of chromosomes, known as X-chromosomes and acting as female-modifiers, and the genes carried in all the other chromosomes which act as male-modifiers. Females have two X-chromosomes and males have one; in the female the female-modifiers in the two X-chromosomes dominate over the male-modifiers in the remaining chromosomes; in the male, the male-modifiers dominate over the female-modifiers in the single X-chromosome.

At the formation of the reproductive cells, all the eggs have one

X-chromosome each, while of the sperm half the number have one X-chromosome and half have no X-chromosome. Depending on the chance whether an egg is fertilized by a sperm bearing one X-chromosome or no X-chromosome, the resulting organism will be female (with two X-chromosomes) or male (with one X-chromosome) respectively.

This mechanism, in which the female has two and the male one X-chromosome, has been found to apply also to many vertebrates including mammals. In the butterflies and moths (as shown by the equally conclusive experiments of R. Goldschmidt) and birds, however, the mechanism is found to be exactly reversed, the female having one X-chromosome (and therefore laying two kinds of eggs, one with one X-chromosome and one with none), and the male two X-chromosomes (and therefore producing sperm of which all have one X-chromosome).

In the higher vertebrates, the birds and the mammals, while the primary determination of sex is effected by the number of X-chromosomes which the fertilized egg comes to contain, one of the early results of development is the production of a gland which secretes a sex-hormone (either male or female as the case may be) which functions not only in controlling the development and differentiation of the sexual characters, but also in maintaining them constant during the life of the animal, thereby avoiding the production of intersexes save in some exceptional cases. In these animals, therefore, the genetic mechanism of sex-determination is backed by a humoral or hormonal mechanism which ensures sex-constancy. In other forms, however, such as the worm *Bonellia*, as F. Baltzer has shown, there is no genetic mechanism of sex-determination at all, but only a humoral mechanism. The young worm develops into a female unless it happens to settle on an existing female, in which case a hormone produced by the female induces the young worm to develop into a male. In other forms again, such as the limpet *Crepidula* studied by J. H. Orton, the production of intersexes by the transformation of one sex into the other (which is carefully guarded against in birds and mammals) forms part of the normal life-cycle. The young limpet settling on a female becomes a male, but in time it transforms itself into a female and receives other young limpets which it causes to develop into males.

Embryology

The study of embryology was begun on a descriptive and comparative basis in the 1820s by K. E. von Baer, and knowledge was gradually acquired of the shapes through which the embryos of certain common animals passed in their development from the egg into the adult. It had been realized for some time that there appeared to be a parallelism between the types of animals that could be arranged in increasing order of complexity along the 'Scale of Beings', and the sequence of embryonic developmental states. The advent of the theory of evolution added great zest to embryological studies, because the old Scale of Beings was then taken to be an ancestral tree and it was hoped that knowledge of the development of animals (or *ontogeny*) would reveal information concerning their evolutionary (or *phylogenetic*) descent.

In many fields, results of great value were obtained. The affinities of certain animals, in which the adult stages are highly specialized or parasitic and degenerate, became obvious when their embryonic or larval stages were found to resemble comparable stages of other less specialized forms. Great strides were also made in comparative anatomy, particularly at the hands of F. M. Balfour, J. W. Jenkinson, J. P. Hill and E. S. Goodrich, by the provision of developmental data with which to interpret the complicated adult structures.

The Error of 'Recapitulation'

Unfortunately, by the exertions of Ernst Haeckel, enthusiasm outstripped judgment. In 1866 he asserted that the series of embryonic stages through which an animal passes in its ontogenetic development and the series of adult forms which represent its evolutionary or phylogenetic ancestry are causally related. This assertion he raised to the level of a 'fundamental biogenetic law', expressed by saying that 'phylogeny is the mechanical cause of ontogeny' and that an animal in its development from the egg 'recapitulates' in its embryonic stages the successive adult forms of its evolutionary history, and thereby he did a great and regrettable dis-service to science.

In the first place, the evidence for the theory of recapitulation is faulty. In some cases it is even necessary to reverse the stratigraphical order of appearance of certain fossils in order to force their series into conformity with the sequence of shapes which the theory would require. Next, it is possible in many cases to show that the embryonic stages of a descendant animal do not represent the *adult* stages of the ancestral forms, but the *embryonic* stages of those ancestral forms. What the study of the embryology of different animals shows is not *recapitulation* (i.e. an abridged and condensed repetition of the ancestral history) but *repetition* of some of the corresponding embryonic stages of the developmental histories of the ancestors.

If the theory of recapitulation were true, it would be necessary to imagine that all evolutionary novelties could only become incorporated in the final stages of the life-histories of a lineage of animals, i.e. in the adult, to sink successively into earlier and earlier stages of ontogenetic development in the descendants. But there is good evidence not only that evolutionary novelties have arisen in early stages of development but that such novelties as these have been of the greatest importance, since the new types that have resulted from them have dropped their old adult forms, and represent major evolutionary advances which have radiated out in many directions to form large new groups. This principle, called *paedomorphosis* (evolutionary rejuvenation) by W. Garstang and G. R. de Beer, appears to have been operative in the production of the most successful of the groups of animals, the insects and the vertebrates, and also in the origin of man.

The converse process of successively increasing complexity and specialization of the adult forms of a lineage by adding small evolutionary novelties to the terminal stages of the life-history, called *gerontomorphosis* (evolutionary senescence), which is the only condition to which the notion of recapitulation can be applied, leads to the production of small categories of animals, species and varieties, and, by refinement of adaptation as well as forfeiture of power to evolve further, exposes them to extinction. The very notion of paedomorphic evolution would have been impossible under the dogma of the theory of recapitulation.

But the theory of recapitulation has had yet another baneful influence on the progress of biology. If it had been true, as Haeckel asserted, that the events of embryonic development were the causal results of aeons of phylogenetic evolution, then there would have been no need to think of investigating the embryology of an animal to see if the structures present at any stage might owe their existence and formation to other structures or conditions present at earlier stages. In fact, the theory of recapitulation delayed the introduction of experimental methods of embryology. And when such methods were eventually introduced, by W. His in 1874 and W. Roux in 1885, they were at first resisted as merely a waste of time.

Experimental Embryology

Experimental embryology was soon able to assert itself as one of the most valuable methods of biological study. While such men as H. Driesch showed that parts of a sea-urchin egg which would normally have given rise only to a half or a quarter of an embryo can, if separated and isolated, be made to develop into miniature whole embryos of half or quarter size, other investigators such as T. H. Morgan and E. G. Conklin showed that in other animals, removal of part of the egg results in permanent loss of one structure or another. Eventually it became clear that these apparently contradictory results were but the expression of differences in the speeds at which certain processes are carried on. In the egg, there are produced certain chemically and qualitatively differentiated 'organ-forming substances' which determine the fate of the portions of the embryo in which they are situated to undergo development along determined lines and to give rise to some definite portion of the organism. If these organ-forming substances are produced rapidly and early, any mutilation of the egg will result in permanent loss and absence of certain parts of the animal. Animals which develop in this manner are said to have *mosaic-eggs*, since the loss of any piece of the mosaic spoils the pattern. If, on the other hand, the organ-forming substances are formed slowly and late, the egg can be divided into two or four, and each of these portions can *regulate* and produce a perfect though miniature organism. Such animals are said to have *regulation-eggs*.

The power of regulation just referred to was, for a time, regarded as a stumbling-block to the possibility of mechanistic interpretation of embryonic development, and led Driesch to formulate his theory of entelechies, which lies outside the realm of science. The solution of the difficulty was achieved by C. M. Child with his theory of axial gradients, invoking the view that the production of qualitative and chemical determinations (i.e. organ-forming substances) in an embryo is the result of a previous quantitative (physical, non-chemical) and graded differential along the main axis of the egg. The axial gradient itself is induced in the egg by the impinging upon it of external stimuli, such as maximum oxygen-supply, light, etc., and the 'high' end of the gradient becomes the anterior end of the embryo. Provided that mutilation of the egg or its division occurs before the quantitative axial gradient has been translated at all its levels into the resultant chemical and qualitative differentiation of organ-forming substances, regulation can occur since that which was lost was not yet qualitatively determined, and, in the portion of the egg which remains, the axial gradient re-organizes the living matter subjected to its influence as if it were a whole egg.

Vast series of experiments performed by Ross G. Harrison have shown that the principle of regulation continues to apply locally to the various regional determinations in the embryo. For example, in newts a certain zone of the side wall of the body becomes quali-tatively determined to produce a limb. But there is at first no precise determination of which parts of the zone, or *limb-field* as it is called, will produce which parts of the limb, and at this stage regulation within the limb-field is still possible. One limb-field can be divided and made to develop into two perfect limbs; or two limb-fields may be spliced together and made to give rise to one limb. At the next stage, as P. D. F. Murray showed in the chick, sub-fields are qualitatively determined to give rise to thigh, shank, ankle and foot; but again, within each sub-field, regulation is still possible. Ultimately, the various parts of the limb are precisely determined.

The course of development is therefore seen to be a succession of qualitative restrictions imposed on the developmental possibilities of the different parts of the egg. Beginning with a state capable of complete regulation, the egg as a whole loses this totipotency when the various organ-fields are qualitatively determined all over it. For a time each of these fields retains its limited power of regulation, which is lost when the sub-fields are determined; and finally all power of regulation is lost.

Little is yet known concerning the precise nature of the physical processes invoked in the function of the axial gradients, or of the manner in which qualitative chemical determination of the various fields is brought about. On the other hand, the experimental investigations of H. Spemann and his school on newts, and of C. H. Waddington on chicks, have yielded much information concerning the property which certain portions of the embryo possess of inducing and forcing neighbouring zones of the egg to undergo particular types of differentiation. For example, the cells which form the roof of the gut-cavity have the capacity of inducing the immediately overlying epidermis to develop into the spinal cord and brain; the eye-cup has the power of inducing the immediately overlying skin to form a lens and then to become transparent. These acts of induction have been shown, at least in part, to depend on the diffusion, out of the inductor-structure, of a substance, the chemical composition of which is not yet known, which evokes the reaction out of the induced tissue.

Inheritance and Acquisition of Characters

While the internal events of the developing embryo are sufficiently complicated in themselves, it is clear that the results of the reactions on one another of the various parts which constitute the causal chain of developmental events are constantly conditioned by the external environment. For example, the development of the paired eyes in a fish or newt embryo presupposes that the water in which the development is proceeding is of normal composition; should there be an excess of certain ions, such as lithium or magnesium, the embryo will develop not two paired eyes but one median cyclopic eye. The principle illustrated by this fact is of fundamental importance for the understanding not only of embryology but also of genetics. The characters which an embryo produces during its development are in some way controlled as regards their ultimate quality by the genes which the embryo has inherited from its parents. But the embryo will produce no characters, or very abnormal ones, in its development if the external conditions to which it is exposed are not normal. The various events of development are responses by the embryo to the complex set of stimuli to which it is exposed both from within and from without.

These considerations enable an answer to be given to the old and vexed question whether so-called 'acquired characters' can be inherited. In the first place, the question is illogically framed, since there is no character of any organism which is not both *inherited* (i.e. dependent on genes inherited from the parents) and *acquired* (i.e. evoked as a response to the stimuli of the environment). Secondly, it is clear that the notion of the length of time necessary for an acquired character to become 'fixed' in inheritance is completely illusory. Few characters can be regarded as so firmly 'fixed' in inheritance as the possession of paired eyes in vertebrates; they are known to have been possessed for over three hundred million years. Nevertheless it is sufficient today to add a pinch of salts to the water in which the development is taking place to show that these aeons of time have not rendered the development of the paired eyes independent of the action of the environment.

The old and fallacious question, "Can acquired characters be inherited?", is now seen to be meaningless, but what its propounders really had in mind was the question: Can a character, the production of which can be shown to be dependent on certain conditions of environment or habit, become incorporated in the genetic constitution of a lineage so that it will be produced in subsequent individuals without the conditions of environment or habit which originally called it forth? In spite of countless attempts to demonstrate the possibility of this occurrence, which involves the question of the induction of adaptive mutation, either directly by the environment (direct induction) or by a habit of life of the organism (somatic induction), no evidence has been produced which would enable this possibility to be accepted.

The Relation of Genetics to Embryology

When the unknown writer of the Hippocratic work *Peri Gones* and Darwin put forward theories of particles, or *pangens*, which were supposed to come from all parts of the body of the parent and, being transmitted to the offspring, moulded it so that the parts of its body resembled those of its parent, two different problems were combined and confused—the resemblance between offspring and parent, which is now recognized as the field of study of *genetics*, and the production of the adult body of the offspring out of something very different, the fertilized egg, a field of study which is now recognized as *embryology*. While these two subjects, genetics and embryology, must be carefully distinguished since no successful analysis of either of them can be made if they are confused, they nevertheless bear very close relations to one another. It has been mentioned that the effects produced by the genes are conditioned by the action of the other genes and by the external environment. Similarly, the processes of embryonic development are conditioned by other events which take place in the organism and by the external environment.

It would be premature to assert that the genes which control the heredity of the characters of an organism are themselves also active as the prime internal factors of development responsible for its differentiation out of the egg; nevertheless it is already clear that the manner in which genes exert their effects is by modifying the developmental production of the character, organs, structure, or colour by which they control. One manner in which this can be brought about has been shown by Goldschmidt and by J. S. Huxley and E. B. Ford to be by means of modification of the speeds at which the processes of differentiation are carried out. For instance, a slow rate of deposition of pigment in the eye results in the appearance of red; a fast rate produces the appearance of black.

In addition, much attention has been given to the study of comparisons between the events of development of organisms which differ in a given known gene. Such studies have revealed valuable information of the 'embryology' of genetic control. H. Grüneberg has shown that a gene in rats which brings out stoppage of growth and premature death does so in the following manner. The primary effect of the gene is to thicken the cartilage of the ribs and of the windpipe, with the result that the thoracic cage is rigid and resists the action of the respiratory muscles, and the cavity of the windpipe is narrowed. This leads to emphysema of the lungs, compensatory hypertrophy of the heart, blockage of the nostrils, inability to suckle, suffocation and death. Behind the simple phrase " a gene controlling this, that, or the other character", there is a complex embryological history to unravel.

Preformation and Epigenesis

Since the days of Aristotle, a controversy has engaged the minds of philosophers concerning the essential nature of what is happening during embryonic development. Is there a new creation of differentiation and complexity when the embryo and adult are produced from an egg; an act of *epigenesis*? Or is development (as the word etymologically implies) an unfolding and rendering manifest of a pre-existing complexity: *preformation*? The difficulty of conceiving how a rational and mechanistic explanation could be given of epigenesis was responsible for the adoption during the seventeenth and eighteenth centuries of a preference for the preformationist theories of development. They soon fell into disrepute. In addition to the farcical controversy on the question whether the future man was 'preformed' in the egg or in the sperm, there was the absurdity of the logical consequences of preformation: for if the next generation is already preformed in the egg, the subsequent generations must also be preformed in what will ultimately become their eggs, and so on *ad infinitum*.

The experimental production of chick-monsters by interfering with their incubation had in the 1820s enabled Geoffroy de St. Hilaire to conclude that since these monstrosities cannot have been preformed, there is no reason to assume that normal chicks can have been preformed, either. All the subsequent work on the regulation of embryos or parts of embryos to undergo normal development in spite of the loss of part of the living matter, confirms the view that the production of an embryo and an adult out of an egg is an act of epigenesis. On the other hand, the bounds of developmental possibility which are set by the genetic constitution of the individual and which it has inherited from its parents, may be regarded as a sort of genetic 'preformation' of potentialities, affecting the individual in question only, and which may or may not be realized according to the success or failure of its epigenetic development.

Conclusion

A century ago, nothing whatever was known of the very elements of genetics, the determination of sex, or of the possibility of application of causal analysis to embryology. Today, all these aspects of biology are firmly grounded on objective and experimental evidence, and the field of promising research is open so wide that the extent of scientific progress may be measured not only by how much is known now compared with then, but also by the vastly greater scope of possible discovery that has been revealed.

CHAPTER XV

PHYSIOLOGY AND HISTOLOGY

by K. J. FRANKLIN, D.M., D.SC., F.R.C.P.

I. Physiology

THE word 'physiology' has had a variety of connotations during the course of its long history, and even now it is undergoing further change as its subdivisions, because of their increasing scope, evolve into separate disciplines. Its purely biological connotation dates from the sixteenth century, and it became more or less limited to this in the eighteenth century. In this Chapter *animal* physiology alone will be discussed; for an account of the progress of plant physiology the reader must seek elsewhere.

In 1851, the former included not only knowledge of the activities of the living organism and of its constituent parts, but also, as legacies from the distant and recent past, variable amounts of comparative anatomy, embryology, and microscopic anatomy. Physiological chemistry, on the other hand, was still in its early stages and its findings were, as a rule, integrated with the other aspects of the subject.

The teaching of physiology was at a very elementary stage of development and practical classes for students were virtually non-existent. In addition, the association of anatomy or of general anatomy (the study of tissue structure) with physiology in the titles of professorships still obtained in most countries. Further, the total number of professional physiologists was very small, physiological societies had not yet come into being, there were only one or two special journals and those foreign ones, and research was prosecuted by a somewhat limited number of enthusiasts in various countries, our own not being particularly prominent in this respect in the year under consideration. Finally, teleological explanation of ascertained facts, i.e. a feeling that definite purposes must be assigned to the various activities of the body or of its constituent parts, was still prevalent. In other words, physiologists continued to concern themselves quite largely with the final cause, as well as with the efficient cause, of phenomena.

In so far as the titles of chairs in Great Britain were concerned, the association of general anatomy with physiology ended in 1874, when Burdon-Sanderson succeeded Sharpey at University College, London. Sharpey had been professor of general anatomy and physiology, Burdon-Sanderson was the first professor of physiology *per se*. This did not mean the dropping of histology, but merely the merging of that study into physiology, which was now beginning to have an experimental side of importance, as well as a not inconsiderable chemical side. After University College had given the lead, chairs of physiology were founded at Oxford (1878), Cambridge (1883), and elsewhere, and the instruction of students began to be established upon more modern lines. In some schools histology has remained with physiology, in others it has become part of anatomy.

Embryology and Biochemistry

Exactly when embryology became a part of structural rather than functional teaching is more difficult to ascertain, but it did become thus divorced from physiology in all medical schools in the period under review, presumably because in its earlier stages it was of value in explaining the origin of structure rather than that of function. In the last two or three decades, through the work of the late Sir Joseph Barcroft and others, it has acquired considerable physiological significance but, in so far as teaching is concerned, it is likely for the present to remain with anatomy.

Later than embryology, but for a different reason, physiological chemistry split off, as biochemistry, from physiology. In this case separation was caused by the phenomenal growth of the chemical side, and the consequent need for separate staffing and accommodation. The change began on the Continent long before it did in Great Britain, where the first chair of biochemistry—a research one at Liverpool—was not occupied until 1902, and the first ordinary one—that at Cambridge—not until 1914, though since then many others have been created. In more recent years there have been signs in some quarters that biophysics is about to split off in somewhat similar fashion, but the tendency is at present fairly localized.

Such diminutions in the scope of physiology must be regarded as inevitable outcomes of the increase in knowledge which has resulted, and is still resulting, from the application of the physical sciences to biological studies. It means that no physiologist of 1951 can have the all-round outlook on the subject which was possible to his predecessor of a century ago, let alone the leisure which that predecessor had for cultural pursuits outside his professional work. In other words, the physiologist of today must specialize in so far as his research is concerned, and even in his teaching he draws much less on personal experience or else limits his range.

There is much other evidence of the ever-increasing complexity of the subject. The first journals devoted to it were a French one and a German one, founded in 1821 and 1834 respectively; they were not followed until 1878 by our own *Journal of Physiology*. Nowadays, there are numerous journals dealing with the science and its subdivisions, and the number of papers published each year is grossly understated by the word 'legion'.

The growth of the Physiological Society, our national one, shows a similar trend. It was founded in London in 1876 for a maximum of forty members. By the third decade of the present century that number was much increased but each member still knew all the others and had some acquaintance with their researches. Now the number is far higher, and it is doubtful if any one of them knows all the rest by name; in addition, there is a long list of applicants desirous of joining the Society.

International Aspects

The story of the International Congresses of Physiologists illustrates the same point. At the first meeting, held at Basel in 1889, there were under 130 members and everyone was able to attend all the sessions. At recent Congresses, despite close scrutiny of qualifications for membership, one to two thousand persons have been present, and several sessions have had to proceed simultaneously in order to deal with the scientific business.

Mention of International Congresses leads, not unnaturally, to some word about the interesting way in which different nations have successively been in the vanguard of progress. It is probably correct to say that France led in the early part of the nineteenth century and continued to be eminent, through such men as Flourens, Claude Bernard and Paul Bert, until well past the middle of that century. Overlapping this period of French activity was the rapid rise of the science among the German-speaking countries, when men such as Johannes Müller, Brücke, Carl Ludwig and du Bois-Reymond, to name but a few, wrested the palm from France. After this phase had been under way for some years, British physiology similarly began to become strong, and over the turn of the century and during the five decades of the new one its outstanding contributions have received universal recognition. Somewhat later in onset than the British, there has been the emergence of North American physiology in the grand manner, and also on a scale with which the old-world countries have not been able to compete.

Compression of the long story, however, is somewhat unfair to the smaller countries, many of which have had as high a proportion of good physiologists as the larger ones in relation to their populations. If, therefore, later in this account more detailed mention is made of our own British contributions, it is with the qualification that 'The great republic of medicine,' of which physiology is a part, 'knows and has known no national boundaries' (OSLER).

The Objective

The objective of physiologists is to acquire as complete as possible a knowledge of the normal functioning of the organism and its constituent parts, at all stages, and under various environmental conditions. To attain this objective, they have observed and recorded activity in the intact and in the operated subject, the operative procedures exposing parts otherwise inaccessible to study, or limiting the biological variables, or both. Between 1851 and the present time, the study of individual parts removed from the body, and kept at appropriate temperature in appropriate media, has been added to the earlier methods; it enables one to determine the potential activities of the part and is a valuable prelude or adjunct to study of the animal as a whole.

Though certain individuals, such as William Harvey, had made very notable contributions much earlier, it is not unjust to attribute the effective beginning of modern experimental physiology.

P

i.e. of operative work on animals, to the efforts of a small number of workers, mainly French ones, in the early part of the nineteenth century. In its development after 1851 it has been radically affected, just as human surgery has been, by the introduction of anaesthetics (first general, then local), antisepsis, asepsis, the use of antibiotics, of coagulants, and so forth. Since 1876, in so far as Great Britain is concerned, its practice has been restricted by law to holders of Home Office licences, working in authorized institutions.

Instrumental Aids

The range of observational power of the physiologist has been increased in various ways, some of which came in before 1851, though many more did so after that date. Apart from advances in microscopy, to be described below, we can cite as an example the introduction of the ophthalmoscope by Helmholtz in 1851. Much more significant, however, than even that notable advance was the discovery of X-rays by Röntgen in 1895. For radiography enables one to see, without operative interference, many of the internal activities of the body, while cineradiography enables a moving film record to be secured of them. For the study of the circulatory, respiratory, and urinary systems in particular, the invention, from about 1930 onwards, of innocuous media which can be introduced to make internal parts opaque to X-rays, has been an invaluable adjunct to the radiographic techniques mentioned.

Apart from such artificial increases in his range of vision, the physiologist has profited greatly by the introduction of photographic recording methods, used either *per se* or in combination with other techniques, in which latter case they are not the major feature though remaining one of importance. When used *per se*, photography produces objective records which convince those who are averse to accepting personal visual impressions. All the forms of photography, including all the devices and kinds of cinematography, have been brought into service, and have proved of great, often of very great, value to the physiologist.

As an adjunct to other techniques, at all events in research, photography has largely replaced the recording method, first introduced by Carl Ludwig in 1847, in which a tracing was made by a pointer writing on a smoked paper carried on a revolving drum. We may mention two cases, which are at the same time interesting instances of the technical developments exhibited by physiology along other lines. The first is the introduction by Sir Charles Sherrington, about thirty years ago, of an optical frictionless myograph to replace the somewhat cumbersome apparatus previously employed in neuromuscular studies. The stage had been reached when practically all that could be determined with the old apparatus had been done, and a more exact technique, free from the criticisms that could be levelled against its predecessor, had to be invented. The optical devices cut out the use of levers, and the photographic recording satisfied the other requirements. The second example is the introduction by Gasser and Erlanger, about the same time, of the cathode ray oscillograph as a means of detecting the very small variations in electrical potential which accompany activity in nerves. Their technique and its modifications began an important new phase in physiological research, for it is a general principle that a part in activity is electro-negative to a part at rest, and, provided suitable amplification of the change is made possible, the results of the activity can be observed on the fluorescent screen or photographically recorded. In many directions new studies have been made possible by such means; in others the time taken to reach conclusions has been reduced to a hundredth of its previous length. Among the leading exponents of such work in our own country has been Professor E. D. Adrian, of Cambridge.

If one reads all the literature on a particular part of physiology, one finds that advances in thought are usually gradual even if, for one reason or another, a particular scientist manages to make such an advance widely or universally acceptable. Hence a full appreciation of past work reveals few *abrupt* changes in physiological outlook. It is true that knowledge of the ductless glands and of the vitamins are new chapters introduced into the text-books during the last hundred years, and one can easily stress particular items such as the work of Oliver and Schäfer on the adrenal and pituitary glands in 1894 and 1895, and of Gowland Hopkins on accessory food factors in the first and second decades of the twentieth century. In each such case, however, there were essential preliminary observations; the story of the vitamins, for instance, had its effective beginnings some centuries back, while the name 'insulin' was coined several years before Banting and Best isolated the secretion in therapeutically usable form, i.e. the fact of its production by the pancreas was accepted and technical difficulties alone prevented earlier workers from anticipating the Toronto ones in their achievement.

We should, then, do wrong if by omitting details for lack of space we made the advances of physiology between 1851 and 1951 appear abrupt and spectacular. All that can, and should, be stated in a brief survey such as this is that most branches of the subject were still in their infancy in 1851, but that they have since developed ever more rapidly through the improvements in technique, the increased help from other sciences, the greater number of workers, and other predisposing factors which have been mentioned in the preceding paragraphs.

British Pioneers

We can, however, note some outstanding British contributions to this general advance and thereby fulfil one of the objects of this book, provided we realize that the list given is merely a selection and that many other names of individual physiologists could justifiably be included in it. In the field of neurophysiology, to begin with, there is the pioneer work of David Ferrier from 1873 onwards on the brain, of Gaskell and Langley on the autonomic nervous system, of Sherrington on the integrative action and numerous other features of the nervous system as a whole, of Dale on the humoral transmission of nervous impulses, and of Adrian on peripheral nerve, the physical background of perception, and so forth. Sherrington, Dale, and Adrian's status has been recognized at home by the conferment on them of the rare Order of Merit, and abroad by the award of Nobel Prizes. Sherrington's book, *The Integrative Action of the Nervous System*, published in 1906, is *the* classic of neurophysiology and in its wider aspects is not yet outdated, nearly half a century after its first appearance. That, for a scientific book, is something very remarkable!

In the physiology of respiration we can point with pride to the great achievements of J. S. Haldane and of Joseph Barcroft and their respective colleagues. Knowledge of the physiology of blood, again, has been greatly advanced by Sydney Ringer, Barcroft,

Peters, Roughton and Adair, to name but a few British contributors. Studies of the blood's circulation have owed much to Gaskell, Starling, W. M. Bayliss, Leonard Hill, Thomas Lewis, Dale, and others; of the lymph and its functions to Starling; of muscle physiology to Fletcher and Hopkins and A. V. Hill (the two last shared Nobel Prizes); of renal physiology to Bowman (1842) and in more recent times to Rose Bradford, Starling, Verney, and others. Finally, to close this short list of examples, the names of Bayliss and Starling are permanently associated with work on the hormones and on the alimentary canal.

All the above men have made outstanding contributions to physiological research, and in consequence of their labour and that of many others not specifically mentioned Great Britain's achievement in the science has for many decades ranked extremely high. On the teaching side, which is less frequently mentioned but which is of obvious importance if medicine is to benefit from the advances made in research, British physiology has a fine tradition, stemming from such men as Sharpey, Foster, and Starling, and continuing with such others as Barcroft, Lovatt Evans, Gilding, and Newton.

Human 'Guinea-pigs'

All told, therefore, if our countrymen were comparatively slow in taking to physiology in the earlier part of the nineteenth century, there was something in the subject with which the national genius seemed to be fundamentally *en rapport*. So, once it had begun, British progress was rapid and British contributions to the advance of knowledge were impressive. One feature which some think to be a relatively recent innovation in physiology, namely, researches in which man himself is the main object of study, is not really so, for about half the passages quoted by J. F. Fulton in his fascinating *Readings in the History of Physiology* describe such instances of 'man as his own guinea-pig'. We may, however, note with pride that physiologists such as J. S. Haldane, J. B. S. Haldane, and Joseph Barcroft have not hesitated, on occasion, to expose themselves to considerable personal risks in their zeal to increase knowledge. We should also remember, in this connection, those unnamed men who, though not themselves physiologists, have undergone similar risks for the sake of their fellows in the occupations of peace and war.

We may quote the instance of the submariner who, in going to a particular diving test, had to step over the collapsed bodies of two of his comrades who had just returned from their tests, but was not thereby dissuaded. The other services, and certain civilian occupations from which danger is never absent, can produce similar examples of quiet heroism by which British physiology has been advanced. Limitations of space, however, prevent further reference to this aspect and we must also, for the same reason, conclude this section on physiology and pass to that on histology.

II. Histology[1]

Histology was defined by Quekett in 1852 as 'the science of the minute structure of the organs of animals and plants'; he also wrote that 'in organized beings, nature works out her most secret processes by structures too minute for observation, unless with the assistance of the microscope'. Our immediate problems, therefore, are to discover when this instrument came into practical use in the field of biology, how and when histology became a separate study, to what extent it had progressed by 1851, and what has since then been accomplished.

Early Microscopy

Though simple instruments were in existence earlier, it was the improvement of the microscope (the first known use of this word was in 1625) during the seventeenth century which led to those considerable discoveries in the biological field which are associated with the names of the so-called classical microscopists, Robert Hooke, Nehemiah Grew, Antonij van Leeuwenhoek, Jan Swammerdam, and Marcello Malpighi. For the time, their achievements were numerous and great; on the other hand, the glimpses of organic structure which these men obtained were mostly imperfect and unconnected observations "from which," as Quekett wrote, "it was impossible to educe any of the general laws of formation and development." Also, though the word *cellula* was used by Hooke in his *Micrographia* in 1665, it had not there its later connotation,

[1] I wish to acknowledge the very considerable assistance received from my friend, Mr. K. C. Richardson. I am also indebted to Mr. R. W. Weeks for help and references.—K.J.F.

and the ultimate cellular structure of plants and animals was not, at the time, even remotely envisaged.

In the eighteenth century there were similar isolated, if able, contributions but still no progress towards a science of minute tissue structure, partly because the microscopes were still rather primitive, partly because in the biological field microscopy had to compete with the stronger appeals of gross anatomy, embryology, comparative anatomy, and physiology.

In 1801, however, the French anatomist, Xavier Bichat, in his *Anatomie générale*, focused attention on the tissues rather than on the organs of the body, and thanks to his clarity of exposition re-orientated the outlook of medical scientists in many countries. He did not himself use anything more than simple lenses, so he prepared the way for histology (a name introduced by A. F. J. K. Meyer in 1819) rather than initiated it. Its actual development was due in large measure to the Czech scientist, J. E. Purkyně or Purkinje (1787–1869), who was appointed Professor of Physiology and Pathology at Breslau in 1823, and who in 1832 got hold of a Plössl microscope and began the long series of studies of animal tissues which made his laboratory, in R. Heidenhain's words, 'the cradle of histology'.

A much younger person who was also among the first to apply the microscope extensively to the study of animal tissues was Johannes Müller (1801–58), and it was one of his pupils, Theodor Schwann (1810–82), who in 1839 enunciated the so-called 'cell-theory', according to which "there is one universal principle of development for the elementary parts of organisms, however different . . . this principle being 'the formation of cells'." Previously to 1839 there had been isolated contributions, e.g. the discoveries of the cell nucleus and nucleolus, but the publication of Schwann's work, despite an unwarranted speculation contained in it, seems somehow to have occurred at the right moment, and it promptly catalysed a large number of studies in various countries, so substantially accelerating the advance of histology.

Soon after Schwann's pronouncement, i.e. in 1841, there appeared what was virtually the first text-book on the subject, namely, Henle's *Allgemeine Anatomie*; between 1850 and 1854 came out Kölliker's more specifically titled *Mikroskopische Anatomie*, in which the author stated that the chief contributions to date had been that of Bichat on the theoretical side, and that of Schwann on the side of fact. He then added: "What has been done in this science since Schwann has admittedly been of great importance to physiology and to medicine, and to some extent of great value from a purely scientific aspect, in that much which Schwann merely indicated, or referred briefly to, such as the genesis of the cell, the import of the nucleus, the development of the higher tissues, their chemical relations, etc., has been elaborated. But all this has not amounted to so great a step forward that it constitutes a new epoch."

If, however, this further development had brought no spectacular advances, it had by 1851 resulted in good accounts of the minute structure of most of the body and in interesting suggestions, based upon such structural findings, as to the physiology of various organs. During the succeeding hundred years, histology profited by various technical improvements, optical microscopy became almost as perfect as is theoretically possible, and other forms of microscopy came into active being. We may begin with some of the technical improvements.

Technique of Microscopy

For an organ or tissue to be studied by means of the optical or light microscope, it must usually be cut into thin slices, or 'sections'. In early days the material was often frozen (freezing by carbon dioxide is a more modern development) and cut by a sort of hand microtome. The introduction of techniques for embedding the specimens in paraffin wax (1869) or collodion (1879) preparatory to cutting them was an important advance, and so was the invention of mechanical microtomes. Differential staining of the sections by natural or artificial dyes was likewise a great forward step, while photography was brought into service, first as photomicrography for recording the microscopical images of sections, then much later as cinematography in combination with methods for direct observation of living materials. The behaviour of silver salts, reduced by photographic developers, also appears to have suggested to Simarro (1900) and to Cajal (1903) the fruitful idea of applying the processes of silver impregnation and reduction to the staining of nervous tissue.

Perfection of the microtome, by making possible the cutting of series of sections at even thicknesses, led to reconstructions of such sections at suitable magnifications either graphically or as models in wax, and by such means embryologists like Born (1883) and Peters (1906) were able to follow the growth changes leading to differentiation of tissues and organs, especially at the early embryonic stages for the elucidation of which dissection methods were impracticable.

Tissue Culture, etcetera

Direct inspection of living tissues and organs is often a desirable counterpart to the study of fixed and stained sections, and early on a beginning was made by observations of the blood flow in the exposed mesentery of the anaesthetized frog and elsewhere. It was not, however, until the first decades of the present century, when the technique of tissue culture was developed by Ross Harrison (1907) and Carrel (1912), that real progress was made. Embryonic tissues, they found, could be grown indefinitely apart from the body in suitable nutrient media, and in consequence valuable additions were made to knowledge of cell structure and behaviour. Later, the limitations of this technique (absence of blood and nerve supply and inadequate chemical control of the environment) led to studies of tissues and organs being made within the intact body by such means as the transparent ear chamber described by Sandison (1924) and developed by Clark (1930); researches of this kind have been greatly aided by the quartz-rod illumination technique introduced by Kniseley (1936).

Living cells and tissues show poor contrast among their various components when viewed by transmitted light, and this difficulty was to a limited extent overcome by the introduction of non-toxic, so-called vital dyes and fluorescent dyes. Ranvier (1875) observed that the white blood cells of the frog could engulf particles of carmine and so arose a means of labelling various kinds of phagocytes. Then in 1885 Ehrlich made the very important discovery that living nervous tissue could be stained during life by methylene blue. Purely optical methods (phase-contrast microscopy; see below) for increasing the relative contrasts in living tissues are of comparatively recent development, but the results so far obtained are much more promising than those secured by *intra vitam* staining.

The routine procedures of fixing and staining the lifeless remains of tissues in the form of sections, films, or smears entail distortion, both physical and chemical, which in extreme cases may introduce appearances or artefacts that are quite atypical of the living cell. While morphological research continued throughout the first quarter of the present century to be the major interest of histologists, attempts were made to study the physical properties of unfixed protoplasm and living cells. Barber (1914), Chambers (1922), and Peterfi (1923) developed the technique of microdissection, while Harvey (1932) and Beams (1933) used the ultracentrifuge on living cells. Injection of indicators into individual cells, micro-surgical operations, estimation of capillary blood pressure, and the spatial separation of the components of the nucleus and cytoplasm were achieved by these methods.

Intracellular Localization

The localizing of various substances identified by ordinary chemical analysis was attempted in histological sections by many early cytologists and histologists. Freezing and drying, as Altmann (1890) found, reduces physical and chemical alteration in sections to a minimum, but it was not until 1932 that Gersh produced the prototype of the modern freezing–drying apparatus. For various reasons, however, the application of reagents giving specific, visible reactions for individual substances is still rather limited in its scope. For the detection of non-volatile inorganic radicles the technique of micro-incineration, conceived by Raspail (1833), was developed by Policard (1927) and by Scott (1935). Study of the more complex organic molecules has been advanced by Caspersson's ultra-violet absorption method. Gomori (1939) and Takamatsu (1939) also made the remarkable discovery that some enzymes remain unaffected in tissue section, despite the severe treatment undergone during fixation and embedding; by appropriate means, the presence of these enzymes can be demonstrated in such sections. In general, dyestuffs have not been used much in histochemistry, as few react specifically with individual substances; an outstanding exception is the reaction of decolorized basic fuchsin with desoxyribonucleic acid in nuclei after mild hydrolysis of a paraffin section (Feulgen, 1924).

Radiography, etcetera

X-ray diffraction techniques, ably developed by Astbury, have been applied to biological materials such as keratin, myosin and collagen, and such biophysical methods are obviously to be of importance in histology. Radiography in all its forms (ordinary, stereo-, cine-, and micro-), aided by the injection of radiopaque substances, is another technique of value which comes from the physical side; it has given valuable static and dynamic pictures of the blood vessels, etc., in both dead and living material. Finally, to this brief and incomplete list of technical advances, we must add the demonstration and localization of radioactive isotopes in histological sections. Photographic plates superimposed on such specimens can be developed in the ordinary way, and such 'auto-radiography' has become routine in investigation of the metabolism of iodine by the thyroid gland, of phosphorus in skeletal ossification, and so forth.

It remains to describe the developments which have taken place since 1851 in the light microscope, and to mention certain other instruments which have been developed or have come into being. The light microscope, with the full development of the homogeneous principle in 1878, the introduction of the apochromatic system in 1886, and the reduction in cost of high-performance components, reached, for practical purposes, its maximum theoretical capabilities; its highest reliable magnification range is between 1200 and 1800 diameters. The single objective binocular microscope, the origin of which is usually credited to Riddell (1851), is easier on the eye, while the double objective binocular microscope, for which credit is as a rule assigned to Greenough (1892), has made possible the development of numerous types of dissecting microscopes giving stereoscopic images at moderate magnifications.

Ultra-microscopy and the Electron Microscope

When, by the beginning of the twentieth century, it was clear that no further development in resolving power (the ability of an objective to show detail in an object) was possible by ordinary light and a system of glass lenses, various other means were tried. With the use of ultra-violet light and quartz lenses double the resolution can be expected, but for purely descriptive work ultra-violet microscopy has not been of any great value. About 1900 Zsigmondy and Sudentopf developed the technique of ultramicroscopy in which the object is illuminated from the side and viewed at right angles to the incident illumination. A simpler arrangement known as the dark-field condenser giving so-called dark-ground illumination has been a valuable aid to the observation of living cells and bacteria, and in 1904 Kohler combined dark-field with ultra-violet illumination. The really spectacular advances, however, date from 1932, when Bruche and Johannson published the first images produced by electrons controlled by an aperture lens system, and Knoll and Ruska developed the first magnetic electron microscope, following on the electron diffraction camera designed by G. I. Finch and demonstrated by him early in 1931 at Imperial College, South Kensington

The technique of this new electron microscopy is still in its infancy, and its present efficiency is only about one thousandth of that theoretically possible. It is developing so rapidly and successfully that no one can readily predict how far the study of micro-anatomy may be advanced by it in the immediate future, e.g. even in the relatively short time which will elapse between the writing of this chapter and its publication. The method gives very high magnification, and stereoscopic pictures can be obtained at even 50,000 diameters; the resolution is a hundred times that obtainable with the light microscope; finally, there is extreme flexibility. Disadvantages of the technique include the smallness of the field, and the facts that the studies must be made in vacuo (i.e. the biological specimens are dehydrated), that differential staining is impossible, and that ordinary staining is at present a drawback, though heavy metal staining may in the future prove useful. To conclude this note, we may add that the electron microscope is primarily designed to allow very high resolution down to molecular level and that, in so far as biological material is concerned, it has already shown amazing things, e.g. what are believed to be contractile molecular threads within muscle fibrils.

The Reflecting Microscope

The so-called reflecting microscope is another instrument which needs to be mentioned. It was originally thought of by Amici of Modena in the early nineteenth century, and thereafter improved by Goring (1826) and others, but it is to the skill of C. R. Burch (1943) that we owe the first high-performance instrument. The reflecting microscope has two unique features—it is completely achromatic and it has an unusually long working-distance. The former means that the entire spectrum from far ultra-violet to infra-red can be plotted on the one instrument so that, with certain other developments which may come in the future, the resolving power of the reflecting microscope may some day be of the same order as that of the present electron microscope. The long working-distance, too, is extremely advantageous for certain types of work such as microdissection, and micro-manipulations can now be performed at high magnification on the surfaces of such organs as the spleen, liver, brain and kidney in the living anaesthetized animal. Finally, the use of an immersion component enables the reflecting microscope to be used for phase contrast microscopy. A valuable adjunct to the instrument itself, namely, a new single-control micro-manipulator, has recently been introduced by Barer and Saunders-Singer.

Phase Contrast Microscopy

Phase contrast microscopy was originated by Zernike of Groningen in 1935 and by its means invisible differences in refractivity are converted into visible differences in intensity, so that images of high contrast are formed from transparent objects of low contrast. It is, therefore, of very great value in the biological field, where the transparency of thin layers of living tissues has usually necessitated fixation and staining, with the consequent possibility of artefact. By the use of the new technique, living cells can be studied without staining under excellent optical conditions, and great advances are being made in knowledge of the structure and behaviour of cells and of the actions of physical and chemical agents upon protoplasm.

It is impossible in this short account to give technical explanations of the electron and reflecting microscopes and of phase contrast microscopy—the only thing that has been attempted has been to give some indication of what their introduction has already meant, and may in the future mean, to histology. The first two are to a large extent specialist instruments, but phase contrast microscopy should be of wider use and may prove a strong competitor with ordinary light microscopy—that remains to be seen. At all events, histology is being vigorously catalysed by these innovations, and it is not out of place to recall the remarkable forecast made in 1853 by Kölliker. "If, without pretensions to prescience," he wrote, "it be permitted to speak of the future, this condition of histology" (i.e. its steady but non-spectacular development from 1839) "will last as long as no essential advance is made towards penetrating more deeply into organic structure, and becoming acquainted with those elements of which that which we at present hold to be simple, is composed. If it be possible that the molecules which constitute cell membranes, muscular fibrils, axile fibre of nerves, etc., should be discovered, and the laws of their apposition, and of the alterations which they undergo in the course of the origin, the growth, and the activity of the so-called elementary parts, should be made out, then a new era will commence for Histology, and the discoverer of the law of *cell genesis*, or of a *molecular theory*, will be as much or more celebrated than the originator of the doctrine of the composition of all animal tissues out of cells." We can with confidence state that the new era thus envisaged by Kölliker has indeed begun.

CHAPTER XVI

BIOCHEMISTRY

by F. G. YOUNG, M.A., PH.D., D.SC., F.R.S.

BIOCHEMISTRY may be defined as the study of the structure and function of living matter on a molecular basis. Defined thus, and always in practice, it overlaps on one side the biological sciences in general and physiology in particular and on the other it intermingles with chemistry and even physics. In 1851 biochemistry hardly existed as such, and indeed it is only during the past thirty or forty years that the functional aspects of living systems have begun to yield to the analytical investigations of the biochemist.

The development of modern biochemistry owes much to the audacity of pioneers who, unabashed by the apparent complexity of the phenomena exhibited by living matter, assumed, on the basis of evidence that was admittedly sparse, that 'life is a chemical function' (Lavoisier, 1780). The rapid and spectacular development of organic chemistry in Germany during the first half of the nineteenth century led naturally to the application of the methods of chemistry to the investigation of the nature of living matter, and the stage was set by the publication, during the decade 1840–50 of three works of major importance: *Chemistry in its Application to Agriculture and Physiology* (1840), and *Organic Chemistry in its Application to Physiology and Pathology* (1842), both by the German, Justus von Liebig, and *The Chemical and Physiological Balance of Organic Nature* (1844) by the French chemists J. B. Dumas and J. B. Boussingault.

Early Views

The view developed at this time, which is still held in a general way, may be summarized in the statement that green plants constitute the main synthetic laboratory of living matter, while animals reverse the synthetic processes and thus obtain for their own use the solar energy trapped and stored by the plant, the excreta of animals providing an important starting material for the synthetic processes of the plant. The idea of cyclic processes in nature, involving the activities of both plants and animals, was obviously of the greatest interest to agronomists, and the founding of the Agricultural Experimental Station at Rothamsted, Herts , by Sir John Bennet Lawes in 1843, was symptomatic of this interest. The development of artificial fertilizers, particularly those containing phosphorus, was an early outcome of progress in this direction.

Nevertheless, despite clamant calls to action in the field of biology, the large majority of chemists remained unconcerned. The chemical structure of many of the substances of importance in biological processes was too complex to yield quickly to the analytical processes then available to chemists; moreover the attractions of synthetic organic chemistry began to multiply and the rich prizes that appeared in the shape of synthetic dyes, perfumes, and even fabrics, proved to be too strong. And so, in the first half of the century 1851–1951 the advances in the chemistry of complex substances of biological importance were as meagre as, in the second half, they were spectacular.

In 1828 the artificial production of urea in the laboratory by Liebig and Wohler had dealt a severe blow to those who believed that 'vital forces' were essential for the reproduction of every phenomenon associated with life. Nevertheless such obscurantist views died hard and at intervals, for a hundred years or more after Liebig and Wohler's experiment, many active research workers found it necessary to declaim against such attempted obfuscation. While such attempts to cloud the progress of biochemistry have always been unsuccessful, let it be understood that biochemistry does not claim to be able to explain all the phenomena of life. It claims only, but firmly, to be allowed to use the methods of chemistry and physics in the elucidation of the nature of the structure and activities of living systems without prejudice, until such time as these methods are clearly shown to be inapplicable. This time has certainly not been reached by the middle of the twentieth century.

It will be convenient to discuss the structural interests of biochemistry before considering its functional aspects, although, of course, the two sides cannot properly be completely dissociated.

Structural Biochemistry

Every living system contains complex nitrogenous substances called *proteins*. The molecules of proteins are large and complex and in water they form a colloidal rather than a true solution. In 1862 Thomas Graham showed that the particles of a colloidal solution will not pass through a parchment membrane, while those of less complex substances will easily do so; thus the separation and ultimate purification of colloids became possible. The chemical attack on the structure of proteins, developed by Emil Fischer in Germany in the 1890s, is being pressed with great vigour at the present time, particularly by Chibnall in this country and by Osborn in the U.S.A. A valuable impetus to such analytical work has been recently provided by the development, by Martin and Synge in this country, of elegant chromatographic methods for the detection and estimation of amino-acids, important constituents of proteins.

Proteins are not only of importance in the structure of the living cell; they form an essential constituent of the catalytic systems of living matter—the enzymes; many hormones, toxins and antitoxins are proteins; the physical properties of the blood, the carriage of oxygen by the blood and the utilization of this gas in the tissues of an animal are all processes which depend upon proteins. It is therefore unfortunate that the chemical complexity of proteins is such that their artificial production has not yet been achieved. The synthesis of proteins in the laboratory is one of the most important goals of modern biochemistry.

Carbohydrates or sugars are also important constituents of living systems. The structure and synthesis of simple sugars were elucidated by Emil Fischer over fifty years ago, and since then an important vitamin—ascorbic acid or vitamin C—has been found to be a simple sugar derivative. The synthesis of ascorbic acid was accomplished almost simultaneously by Haworth in this country and by Reichstein in Switzerland in 1933.

In living tissues simple sugars are often joined together to form highly complex polysaccharides of which starch and glycogen (animal starch—Claude Bernard, 1857) are examples. Polysaccharides are of importance in the cement which fixes together the cells of an animal tissue, and also in the toxic and invasive actions of certain bacteria which can infect animal tissues. Polysaccharides have never been completely synthesized in the laboratory although some have been produced artificially apart from living systems under the influence of enzymes (Cori, 1939; Hanes, 1940). Their structure has been the subject of intensive investigation in recent years, and in a few instances has been almost completely determined, though much still remains to be done.

Fatty substances form an important part of the structure of living cells, particularly in relation to the delicate cell membrane through which all material entering and leaving the cell must pass. Simple fats are formed by the combination of glycerol (glycerin) and fatty acids, but in cells, and particularly in those of the brain, complex fats are found which contain phosphorus, sugars and certain basic substances. The work of Thudichum (1860–80) laid the foundations of our knowledge of these complex substances, many of which still defy synthesis. The class of fats also includes carotenoids, of which many plant pigments and vitamin A are important derivatives, and the sterols and steroids, which include vitamins D and certain important hormones. Vitamin E is also to be included in the class of fat-soluble vitamins.

However stable a living structure may appear to be it is, in fact, in a continuous and rapid state of dissolution and reconstruction. The extreme rapidity with which these processes of destruction and rebuilding are simultaneously taking place has been vividly underlined by experiments in which rare isotopes have been incorporated into living tissues. The use of such tracer isotopes has revealed the fact that carbon dioxide, long believed to be only a product of the breakdown of foodstuffs and tissues in the animal body, can be utilized by animal tissues for the building up of complex substances. In this respect animal tissues and those of micro-organisms share what was once believed to be an exclusive function of the tissues of the plant.

Animal Nutrition

That an animal is a heat engine, liberating an amount of energy and heat precisely equivalent *in toto* to that calculated from the heat liberated when the foodstuffs combusted are burned outside the body, was made clear from the pioneer experiments of Voit and

Pettenkofer begun in Germany in 1860. In that year Voit took with him from London to Munich a Thomson calorimeter and thus initiated the science of animal calorimetry, developed by Rubner under Voit's guidance in Germany, and by Atwater, Benedict and others in the U.S.A. The peculiar significance of proteins as animal foodstuffs was emphasized by these investigators, and later the nutritional importance of certain amino-acids—themselves constituents of proteins—was found to be at the root of these phenomena. Thomas and Abderhalden, in the early years of the twentieth century, and more recently Rose, have determined which of the amino-acids the animal body is unable to synthesize and which are therefore essential constituents of a healthy diet. Nevertheless the reasons for the dietary indispensability of some amino-acids and not of others are still unexplained.

Biochemists have long been attracted by the possibility of rearing and maintaining an animal on a purely synthetic diet, and as long ago as 1906, from experiments directed towards this end, Gowland Hopkins deduced that accessory food factors were important for the development and health of the animal body. Such accessory food factors were not among the more common constituents of the diet, and were important, not because of their contribution to the calorie value of the food, but because they served a specific function at that time not understood. Already, a century and a half before, the value of fresh fruits and vegetables in the treatment of scurvy had been adequately demonstrated, and in the period 1890–1900 the classical experiments of Eijkman had indicated the efficacy of rice polishings in the treatment of beri-beri, while a little later Grijns had clearly conceived beri-beri to be the result of a specific dietary deficiency. Although investigations such as these had revealed that particular diseases might result from nutritional deficiencies it was many years before the importance and widespread nature of deficiencies of this type were suspected.

The Vitamins

From 1912 onwards the work of Osborne and Mendel, and of McCollum, in the U.S.A., and of Hopkins, Funk, Drummond and Mellanby in Great Britain, began to reveal the significance of accessory food factors, or *vitamins* as they later came to be called.

In 1925 Bourdillon and his colleagues of the staff of the Medical Research Council first produced an artificial substance with vitamin activity when they irradiated certain sterols with ultra-violet light and so obtained a substance active in preventing or curing rickets. Nearly all the known vitamins have now been obtained in a pure state and synthesized in the laboratory; foremost in synthetic work have been Todd and Heilbron in Great Britain, Williams and Folkers in the U.S.A., Kuhn in Germany and Karrer in Switzerland.

The role of certain of the vitamins in the processes of metabolism—that is the processes whereby substances are broken down or built up in the tissues—have been revealed by the discovery by Euler and Myrbäck in Sweden, by Warburg and Lohmann in Germany, and by Peters in Great Britain, that certain members of the class of vitamins B are constituents of co-enzymes in the animal body. This can best be illustrated by reference to a specific instance.

Vitamin B_2 was isolated in 1933 by Kuhn in Germany and Karrer in Switzerland, as a substance whose deficiency causes dermatitis, eye changes and sometimes anaemia in animals. This vitamin was synthesized by Kuhn and Karrer in 1935 and named riboflavin. In the animal tissues riboflavin is combined with phosphoric acid and the compound so formed constitutes a co-enzyme for certain dehydrogenase enzymes. These enzymes assist the removal of hydrogen from food substances in the tissues—an essential step in the process of oxidation—and pass the hydrogen to other enzymes and co-enzymes whence ultimately the hydrogen is joined with oxygen to form water. The temporary combination of the hydrogen released from the foodstuff with the co-enzyme is an essential step in this process, and since the body cannot itself produce riboflavin to meet its needs, this constituent of the co-enzyme becomes an essential food factor—a vitamin in other words.

The nutritional importance of inorganic substances has been emphasized within the past twenty years by the discovery that minute amounts of certain elements, notably metals, are of great nutritional importance for both plants and animals. In some instances such elements are found to be essential constituents of enzyme systems. The practical importance in nutrition of trace elements, as they are called, has now been widely recognized, both by agriculturalists and by doctors.

Enzymes

Although foodstuffs are relatively stable outside the animal body and can be preserved in air for days or weeks, once they enter the body they are smoothly and rapidly combusted at a temperature only a little above that of the surrounding air. Such processes of combustion within the body are aided or catalysed by proteins called *enzymes*. Enzymes are also responsible for much of the digestion of foodstuff which goes on in the gastro-intestinal tract. Although it was recognized before the nineteenth century that digestive enzymes or ferments could exist and act outside living tissues, it was a matter of controversy for many years during the latter part of the nineteenth century as to whether the metabolic processes which take place in living cells, for example the fermentation of sugar by yeast to give alcohol, could occur apart from the living cell. Pasteur in particular insisted that fermentation was dependent upon intact living cells for its existence (*la vie sans air*). By the preparation of an extract of yeast from which all intact cells had been removed but which nevertheless could bring about the fermentation of sugar, Buchner (1897) settled this question completely, and since that time a very large number of enzymes have been isolated from animal tissues (Northrop).

The characteristics of enzyme action have been intensively investigated and many successful attempts made to reconstruct, by means of enzymes removed from living matter, events which normally take place in the living cell. In fact modern biochemistry is built up very largely around enzymes and enzyme action, and analysis of the events which take place in the multi-enzyme systems (Dixon) of the living cell is a matter of current and intensive study.

Nevertheless, as yet remarkably little is known of the mechanism whereby solar energy is utilized by green plants for the building up of the complex chemical substances essential to the nutrition of animals. Tracer isotopes have been largely used in recent studies of the mechanism of the photosynthetic production of starch from carbon dioxide and water with the release of oxygen, and Robin Hill has demonstrated (1940) the production of oxygen by isolated chloroplasts (the chlorophyll-containing units of green plants) under experimental conditions. Notwithstanding, the field of plant biochemistry is one which is relatively unexplored.

On the other hand, there is a wealth of information concerning the mechanisms involved in the release or transfer of energy in animal tissues. The classical investigations of Voit, Rubner and others in Germany from 1860 onwards were concerned with the total amount of energy released when food is burned in the animal body. The elucidation of this process in terms of known chemical reactions was begun in the early years of the present century. Chittenden and Lusk were active in the U.S.A. at this time, Knoop's classical investigations were in progress in Germany, while the experiments of Fletcher and Gowland Hopkins concerning lactic acid production in muscle were published in 1907.

Energy Release in Muscle

Whether or not a muscle produces lactic acid when it contracts is a question which had been argued for more than fifty years before 1907. In that year Fletcher and Hopkins established that the accumulation of lactic acid in contracting muscle occurs only in the absence of oxygen and that if lactic acid had accumulated in the absence of oxygen the admission of oxygen then led to the prompt disappearance of the lactic acid. These observations were the starting point of work that is still in progress concerning the mechanism of energy release in contracting muscle. The work was taken up by Embden and Meyerhof in Germany and by Parnas in Poland, among many others, and has led to a detailed knowledge of the chemical reactions, catalysed by a series of enzymes, whereby glycogen or glucose breaks down in muscle with the release of energy. Direct measurements of the heat evolved in contracting muscles, made by A. V. Hill from 1911 onwards, were of outstanding value in this connection.

In 1904 Harden and Young began a series of researches, concerning the enzymes in yeast responsible for alcoholic fermentation, which revealed the importance of phosphoric acid derivatives of sugars in this respect. Since then the importance of organic phosphoric acid derivatives in energy exchange in tissue has been again and again emphasized, particularly by Carl and Gerty Cori in the U.S., and by Dorothy Needham in England, and within the last few years Lipmann has put forward the theory that much of the energy released in metabolic processes is stored in the form of high-energy organic phosphoric acid derivatives. Phosphoric acid is indeed a substance of unique importance in biochemistry.

That iron and iron-containing substances are concerned in the carriage of oxygen in the body was recognized in the latter part of the nineteenth century (Hoppe-Seyler), but the specific role of iron derivatives as catalysts in the tissues, foreshadowed by MacMunn in 1886, was not elucidated until the 1920s when the work of Warburg, and more especially of Keilin, was published. The discovery by Keilin of specific iron-containing oxygen catalysts, widespread in living tissues and named by him *cytochromes*, was an event of outstanding importance in this connection.

Hormones

Hormones, which are the products of certain glands in the animal body, are liberated into the blood stream and so are carried to the points of the body at which they act. A minute amount of such a chemical substance elaborated by an inconspicuous gland may be essential for the life and well-being of man or animal alike. The idea of hormones has also been extended within recent years to the plant world.

In 1855 Thomas Addison published an account of a fatal disease in human beings (now called Addison's disease) associated with degeneration or disappearance of the adrenal glands—two small glands each situated above either kidney. Subsequently it was found that surgical excision of the adrenal glands from an animal caused death and many attempts were made to prepare extracts of adrenal tissue the administration of which would prevent the death of animals from which the adrenal glands had been excised. It was not until well on in the present century, however, that attempts of this type began to be successful. (See also Chapter XVII.)

In the meantime there was a growing body of evidence that other glands in the body contained and liberated into the blood stream hormones of various kinds, some of which were essential for life while others were not. The term *hormone* was put forward by Bayliss and Starling in 1902 as a general term for chemical messengers carried by way of the blood stream. The preparation of insulin by Banting and Best in 1922 provided a hormone of exceptional interest, and of great importance in the treatment of human diabetes mellitus, and their work produced a widespread interest in the subject of hormones generally.

Hormones fall into a number of unrelated chemical classes and are to be grouped together only on biological grounds. A few have a relatively simple chemical structure and have been synthesized in the laboratory. The first to be so prepared artificially was adrenaline, which is secreted by the medulla of the adrenal glands—a part which can be removed without causing death. Adrenaline was isolated by Takamine in 1901 and synthesized by Stoltz and others in 1904. The isolation of thyroxine, an iodine-containing hormone, from the thyroid gland was announced by Kendall in 1915, and the substance was synthesized by Harington and Barger in 1926. A number of the hormones are closely related to the sterols, and these hormones are known collectively as *steroids*. The hormones of the adrenal cortex, the absence of which causes Addison's disease and death, are steroids; two important ones are cortisone (Kendall, 1936) and corticosterone (Reichstein, 1937). The sex hormones, responsible for the development of secondary sex characters such as the beard in the male and the breasts in the female, are also steroids (Butenandt, Marrian, Laqueur, Doisy, 1932-6). Most steroid hormones have not been completely synthesized in the laboratory, though equilenin was synthesized by Bachmann (1940) and oestrone by Miescher(1948), but many can be produced artificially from sterols (Ruzicka, Butenandt, Reichstein). In some instances relatively simple chemical substances have been produced with a very high degree of hormone activity (Dodds, 1937).

Other hormones are protein or protein-like in nature. Insulin is a good example of this class, which includes the hormones of the pituitary gland concerned with the control of growth and with the regulation of the activity of other hormone-secreting glands. Protein hormones have not as yet been synthesized.

The term *phytohormone* or *auxin* has been applied to substances produced in the tips of growing plant shoots which diffuse down into the shoot to stimulate growth and therefore elongation of the shoot. The production of these hormones is concerned in the phenomena of geotropism and phototropism of seedlings. The substances responsible were isolated by Kogl in Holland in 1933, and also by Thimann in the U.S.A. The substances isolated were of two chemical types, one complex (true auxins) and not easily obtainable by synthesis, and the other relatively simple chemically (heteroauxin or indolylacetic acid) and easily synthesized chemically. Within the last few years numerous substances related to indolylacetic acid have been found to exert a profound influence upon various aspects of plant development and growth. Thus we now have highly selective weed killers, such as 2:4-D (2.4.-dichlorophenoxyacetic acid), substances for preventing fruit drop, delaying bud development, and for many other purposes.

The question of how hormones, and particularly those of the animal body, act to bring about their profound physiological effects is one that is of great current interest. It is almost a matter of faith with biochemists that hormones must influence ·enzyme systems, either directly or indirectly, in the course of exerting their characteristic activity. Little is known as yet, however, about the interaction of hormones and enzyme systems although Cori in the U.S.A. has recently obtained some significant relevant results.

Comparative Biochemistry

Although the major part of biochemical interest has always lain with the higher animals and plants, the biochemist has always concerned himself with micro-organisms such as yeast, and within the past thirty years or so his interest in micro-organisms has developed apace. The study of microbiological biochemistry began with the great French chemist, Pasteur, whose work on alcoholic fermentation is classical, and was continued by Bertrand and Winogradsky. The subject has been fostered in Great Britain by Raistrick, Fildes and Stephenson within recent years, and is at present a rapidly growing one.

Surprisingly enough, a study of the nutritional requirements of micro-organisms, initiated by Fildes in the third decade of the present century, revealed remarkable similarities, for example with respect to vitamins and amino-acids, between the requirements of many bacteria and those of higher animals. Indeed, we can now say that there is an impressive similarity in the metabolic processes of all forms of living matter, and it may be assumed that they have been elaborated on the basis of a common 'ground-plan'. Such metabolic similarities are often of value in permitting the biochemist to work out fundamental processes in the less complex systems, and then to investigate the possibility that such results are applicable to higher animals. The comparative biochemistry of animals and plants in general is a fruitful study of this type, as also is that of chemical embryology (Joseph Needham).

Viruses

Ultra-filterable viruses are the smallest known agents which exhibit certain of the phenomena associated with life. Thus although they cannot flourish or reproduce apart from living tissues they can multiply rapidly once a living tissue has been successfully invaded. In 1935 Stanley announced the isolation of certain plant viruses in a crystalline state and subsequently the isolated material was shown to be largely, if not entirely in some instances, a member of the special class of complex proteins known as nucleoproteins (Bawden and Pirie). This development has led to much discussion as to whether such material is living or non-living but since the nature of the conclusion depends upon the definition of the words used such discussion appears to be fruitless. The suggestion that such viruses are similar to genes, if not identical with them, and are thus capable of organizing living matter in a fashion characteristic of themselves, is a speculation that awaits further experimental investigation.

Genes in Biochemistry

The relationship between genes and biochemical processes has been vividly illuminated during the past ten to fifteen years, notably by the investigations of Beadle and Tatum in the U.S.A. These collaborators investigated the influence of irradiation with X-rays upon the nutritional requirements of certain micro-organisms, particularly those which required little preformed organic matter and vitamins. Irradiation with X-rays often induces inheritable changes in the characteristics of living organisms, and the changed offspring are known as *mutants*.

The mutants of the micro-organisms investigated by Beadle and Tatum had, in general, more stringent nutritional requirements than had the parent organisms. For instance, the mutants might be unable to grow without the addition of certain vitamins or amino-acids to the culture medium, although the parent strain grew perfectly well in the absence of these amino-acids or vitamins from the culture medium, and indeed could be shown to synthesize them from other substances during the course of growth. Further investigations showed that the mutant organisms had lost certain enzymes capable of assisting the synthesis of the appropriate vitamins or amino-acids, and that the loss of these enzymes was in turn dependent upon the loss of genetic factors.

This experimental production of potential nutritional deficiency indicates possible reasons for the apparently capricious nutritional requirements of man and higher animals, since these may have been determined by the induction of mutations in ancestors at a time when the diet was rich in substances which, for instance, we now call vitamins. A loss of ability to synthesize these substances would then confer no disability on the animal but indeed might give it some advantage in preventing its performing unnecessary metabolic syntheses. It is only in the civilized state that man has been able to dissect his food and thus to reveal and suffer from hitherto latent metabolic deficiencies. The possibility of throwing light on evolutionary processes by means of biochemical investigation of this type is a fascinating one for future development.

These recent experimental developments bear witness to the genius of Archibald Garrod whose classic work on *Inborn Errors of Metabolism*, published nearly fifty years ago, ascribed certain congenital metabolic abnormalities in human beings to an inherited deficiency of the appropriate enzyme system. This was at a time when the idea that metabolic processes in the animal body took place in an orderly sequence of reactions, each catalysed by an appropriate enzyme system, had hardly been conceived by those responsible for its later development.

Chemotherapy

The development and synthesis of drugs capable of destroying bacteria or other organisms which infect higher animals, while leaving the host unaffected, was the dream of the German biochemist Paul Ehrlich in the early days of the present century. Although Ehrlich's name will always be associated with the introduction of salvarsan for the treatment of syphilis, his ideas outran the biochemical knowledge of his time and it is only within the past twenty years that a rational development of chemotherapy has proved to be possible.

The discovery, by the German Domagk in 1935, of the remarkable efficiency of the synthetic drug prontosil for the cure of certain bacterial infections in man was the starting point of much of the modern research in chemotherapy. Prontosil was a dye effective in staining bacteria and thus of acting particularly on the bacterium as opposed to the tissues of the host. Its discovery led to the development in Great Britain of sulphapyridine (M & B, 693), a substance remarkably efficient in pneumonia, and to a whole host of other substances based upon sulphanilamide, the effective portion of the prontosil molecule.

As the result of their investigations of the mechanism of action on bacteria of sulphanilamide and of other drugs, Woods and Fildes put forward, in 1940, the view that the chemotherapeutic agent was sufficiently like essential nutrients of the bacterium to be taken up by the enzymes of the bacterial cell, but sufficiently unlike to cause damage or blocking to the enzyme system and thus to prevent the growth or development of the cell. This fruitful theory has led many investigators in all countries to synthesize chemical substances similar to, but different in detail from, vitamins, amino-acids and other cell nutrients, and to test these for an inhibitory effect on the action of the related nutrient. Many chemical caricatures have been found to exert the expected inhibitory effect, and thus to prevent the development of the organisms requiring the appropriate nutrient.

The view that substances may compete with each other for the possession of enzymes was developed in the 1920s by Stephenson, Quastel, Haldane and others in this country and had a most direct and fruitful development during the recent war in the discovery of BAL by Peters and his colleagues at Oxford. In investigating the possibility of producing an antidote to the war gas Lewisite, Peters deduced that Lewisite exerted its noxious effects by reacting with certain essential groupings in the enzyme molecules in the tissues, and so preventing these enzymes from performing their normal functions. Peters, Stocken and Thompson therefore synthesized simple substances, of chemical constitution similar to that of the reactive groupings of the complex enzymes, with a view to finding one that would react more vigorously with Lewisite than did the enzymes themselves, and would thus protect the enzymes from the action of the Lewisite. In this way they were able to produce a remarkably effective antidote, named *British Anti-Lewisite* (BAL) by American colleagues. Subsequently, and on good theoretical grounds, BAL was found to be an effective antidote to other poisonous substances, and in particular to certain toxic heavy metals. Truly such developments may be regarded as the biochemical equivalent of beating swords into ploughshares.

Fleming discovered the antibacterial activity of penicillin, a product of the metabolic activity of certain moulds, in 1929, but it was not until 1940 that Florey and his colleagues, in a systematic survey of the clinical value of what are now known as natural antibiotics, showed the uniquely valuable properties of penicillin as a therapeutic agent. Since then the development of streptomycin and other antibiotics in the U.S.A. has greatly extended the armament of the physician against bacterial infection, but unfortunately many of these valuable substances are of such a complex nature that their artificial production by chemical synthesis is not as yet a practical measure.

Conclusion

One hundred years ago the activities of living tissues appeared to present an almost insoluble complexity for chemical analysis. Moreover, the chemist was attracted to the easier paths of synthesis, by the methods of organic chemistry, of relatively simple substances of industrial value. How far the claims of industry may be responsible for the relatively poor interest in biochemistry in Great Britain until about fifty years ago is a matter for economists or sociologists to consider. The development of the special methods needed to handle and analyse large and complex molecules, such as proteins and polysaccharides, was a pre-requisite for the development of dynamic biochemistry. The fruits of such a development, gathered in the fields of medicine and of agriculture during the past thirty years, bear full witness of the value to the world of what may sometimes appear to be academic researches with no obvious practical application.

CHAPTER XVII

MEDICINE, SURGERY AND THEIR
SCIENTIFIC DEVELOPMENT

by E. Ashworth Underwood, M.A., B.SC., M.D.

A foreign visitor to the Great Exhibition in 1851 must have been impressed by the richness of the scene which was spread out for his inspection. The magnificent 'crystal palace', the numerous exhibits on industrial methods, the stability and wealth of fashionable London, must have implanted themselves deeply on his mind. If he had strayed into some of the slums of the great city, he would have witnessed revolting scenes; on him they probably did not make this impression, for the conditions in all great cities were then the same.

Neither would the visitor have been impressed if told that he ran a risk of contracting typhoid fever—which affected the houses of even the richest classes—or typhus; or that cholera might reach these shores, thus lessening his chances of escape. If he had had the misfortune to sustain a compound fracture of his leg, it would almost certainly have been amputated. It is true that he would have had a general anaesthetic, given in a rather crude fashion; but after the operation he would have been lucky not to die of hospital gangrene. What we would now term 'an acute abdominal emergency', for immediate operation, did not then exist, for practically no one had opened the abdomen without fatal consequences. None of these possibilities would have scared away the foreign visitor because the social and medical conditions of London were as good as anywhere else. The physicians and surgeons of the Metropolis were noted everywhere for their skill. Yet the death-rate in 1851 for England and Wales was 22.0 per 1000, as against 12.3 for 1947. The difference represents a transition from one age to another; from empirical medicine to preventive medicine, followed by the present age when scientific medicine has come into its own. The last hundred years have transformed medicine from an empirical art to a highly organized and effective science.

The Birth of Preventive Medicine

In 1832 Great Britain had suffered from a devastating epidemic of Asiatic cholera, a disease new to this country. The 31,500 deaths in the country as a whole gave rise to dissatisfaction with the sanitary conditions of the people. The General Board of Health which had been appointed in London became defunct after the epidemic subsided. Meanwhile men such as Southwood Smith, Sir Edwin Chadwick and Lord Shaftesbury emphasized the point that each large community requires a central medical director to improve its social and sanitary circumstances. In 1847 Liverpool appointed William Henry Duncan as its first medical officer of health, and in the following year John Simon became the first medical officer of health of the City of London. In his six years of office before he went to Whitehall, he produced a series of reports which have been models for those who followed.

In the meantime, and again in company with most of Europe, Britain had in 1848–9 experienced another disastrous cholera epidemic, repeated in 1853–4. During these two epidemics there were 73,400 deaths in England and Wales alone. The social conscience was at last fully awake. In 1848 a permanent General Board of Health was appointed and the first Public Health Act was passed. Subsequent legislation, consolidated in the great Public Health Act, 1875, and energetic measures taken to secure effective sewerage, improve the water supply and ameliorate housing conditions and sanitation, established the necessary environment for later measures against infectious diseases.

The Birth of Modern Surgery

The great operators of the early nineteenth century—here and in Europe—were extremely dexterous. Operating without a general anaesthetic, a Liston or a Syme would carry out an amputation through the thigh in an incredibly short time. But they were practically confined to the limbs, the breasts, and certain parts of the face. These limitations were removed by the introduction of general anaesthesia, and by the growth of the antiseptic method of Lister.

The search for a general anaesthetic culminated in W. T. G. Morton's demonstration of the value of ether at Boston on 16th October, 1846, and in Robert Liston's historic demonstration at University College Hospital, London, on 21st December, 1846, of its use in major surgical operations. In 1847 Sir James Young Simpson introduced chloroform for women in labour, and for many years ether and chloroform were widely used. Within a few months the necessity for speed in operating disappeared, and the complete relaxation of muscles given by general anaesthesia permitted abdominal exploration and the subsequent triumphs of surgery.

These advances would have been of little avail but for the genius of Joseph Lister (later Baron Lister). Appointed at the age of thirty-three to the Chair of Surgery at Glasgow, he found that in his wards at the Royal Infirmary the death-rate from hospital sepsis and gangrene, especially after amputations, was very high. After preliminary scientific work Lister turned to the causes of suppuration. He grasped the importance of Pasteur's discoveries in relation to fermentation, appreciated that suppuration was due to organized living bodies, believed that they were introduced to the wound from the air, and experimented with carbolic acid as an antiseptic (1865). In 1867 he published a paper which established his 'antiseptic method'.

By this paper Lister revolutionized surgery. On the Continent of Europe, especially in Germany, his methods were followed with great success. But in England he was for long a prophet without honour. Lister did a vast amount of work on wound dressings, on the relation of bacteria to inflammation and on materials suitable for ligatures. His system developed into the 'aseptic system' in which germs are not killed in the wound but prevented from gaining access to it. This system is in use in civilian surgery today, but the highly infected wounds received in battle made the surgeon turn back again in war to Lister's original principles.

The Rise of Bacteriology

Although bacteria had been seen by Leeuwenhoek late in the seventeenth century, there was no general appreciation of their effects until Pasteur's work on lactic-acid fermentation and on spontaneous generation around the 1860s. The researches of

Ferdinand Cohn and Carl Weigert put the infant science on a sound basis. Bacteriological technique was standardized by Robert Koch, who in addition was credited with the discovery of the organisms of eleven known diseases, among these being the tubercle bacillus (1882) and the organism causing cholera (1883). In 1877 he had demonstrated the life-cycle of the anthrax bacillus. By the end of the century the specific organisms of many infectious diseases had been discovered.

This prodigious growth of the new science bred a spirit of confidence. The discovery of the organisms causing all known infectious diseases seemed a distinct possibility, and a solution of the problem of cancer on these lines was anticipated. It was not long before there was a check to such vain imaginings. Some common diseases, such as measles and chicken-pox, refused to give up their secret. Smallpox gave many false scents—and still constitutes a problem. The influenza bacillus, isolated from cases of that disease and described as the cause, failed to satisfy the primary laws of bacteriology—the three laws which are commonly known as 'Koch's postulates'. Though present in cases of influenza, this bacillus was not the cause.

It was discovered that the virus of 'mosaic disease' of the tobacco plant passes through the filters commonly used to intercept bacteria. These bacteriological filters are made of unglazed porcelain or similar materials, and their pores are so small that they prevent the passage of all the bacteria which were known at that time. This tobacco virus was not only 'filterable'; it was 'ultramicroscopic'. It was soon shown that many infectious diseases affecting the human organism are caused by such ultramicroscopic viruses. Very powerful types of ultramicroscope, and the method of cultivating the virus by tissue-culture, are now at last solving these difficult problems.

Progress in bacteriology was checked when it came to be realized recently that, while there could be no disease without the causative organism, the presence of the germ did not necessarily imply infection and onset of the disease. In other words, disease was the resultant of forces acting from opposite sides: on the one hand, the virulence of the infecting organism; on the other, the immunity of the receiving host, together with other factors, environmental, seasonal, nutritional, etc., which together constituted a formidable epidemiological study.

For example, the last seventy years have seen various theories regarding the mode of transmission of respiratory diseases. The Munich school of hygiene under Pettenkofer held that dirt and dust, objects touched by the patient, and the air itself carried the germs to healthy individuals. Early in this century opinion changed under the influence of Flügge in Germany and Chapin in America to the view that direct contact, or something approaching it, was necessary. Fever hospital practice and the form of fever hospitals were based on these latter theories. During the recent war, however, it was shown that certain respiratory infections could indeed be 'airborne' for considerable distances. Although new methods of preventing such secondary infection are in use in hospitals, the problem has not yet been solved.

Artificial Immunity

Strictly speaking, immunology was born over fifty years earlier than bacteriology, since Edward Jenner introduced in 1798 his method of vaccination against smallpox. In 1880 Pasteur described his method of using attenuated living cultures for the prevention of fowl cholera, and later his methods for the prevention of anthrax (1881) and of rabies (1885). The method used today for preventing rabies is virtually as Pasteur left it. Treatment by the introduction of immune substances has developed from the discovery of diphtheria antitoxin by von Behring in 1890, and from the classic researches of Paul Ehrlich on the standardization of such antitoxins. Diphtheria antitoxin was first used for the treatment of diphtheria in the human in 1894, and there soon followed other antitoxins, used in the treatment of tetanus, bacillary dysentery, cerebro-spinal fever and other conditions. The use of these sera is the science of immunotherapy. Side by side with its growth has been that of the methods of active immunization.

About the turn of the century Sir Almroth Wright introduced his typhoid vaccine, and it received a try-out in the last phase of the South African war, during which the British lost more men from typhoid fever than from enemy bullets. The first full-scale use of this vaccine was in the First World War, when it achieved a singular and unequivocal triumph. From this work developed the method of using autogenous vaccines for treatment.

About 1890 it was shown in Germany that diphtheria toxin, mixed with an appropriate amount of specific antitoxin, could be used in the human to produce an 'active' immunity to natural infection with the diphtheria bacillus. This type of immunity implies that the individual is able to respond to the stimulus of infection by suddenly producing sufficient antitoxin to overcome the infection. In 1913 Bela Schick of Vienna introduced his simple test for suscepti- bility to diphtheria, and immunization on a wide scale became a practicable proposition, first followed whole-heartedly in America under the guidance of W. H. Park. Dangers associated with the use of pure toxin in the mixture were overcome by substituting for it a modification known as toxoid. This led to the development of the even more safe and powerful diphtheria prophylactics of the present day. By these methods diphtheria has been completely eradicated from certain Canadian cities, and in the United States, France and Great Britain the dangers of the disease are markedly lessened. Artificial immunization against other infectious diseases is being closely studied but many technical difficulties are involved.

The New Pathology

For centuries scientists had debated the question whether diseases started in the tissues of the body or in its juices (blood, lymph, etc.), or *humours* as they had been called. Towards the middle of the nineteenth century opinion had hardened in favour of the tissues. In 1857 Rudolf Virchow demonstrated that the ulti- mate starting-point of disease is the living cell. It is therefore in the cells of the tissues that characteristic disease changes must be sought. This doctrine of 'cellular pathology' has had revolutionary results. Allied to it was the study of minute changes in the normal cell—especially those changes in the chromatin network which lead to the formation of chromosomes, constant in number for each species. It is to these studies that we owe the scientific theory of heredity, which itself has many medical applications.

The General Physician

The new concepts which were being introduced into the practice of medicine were making it more difficult for any one man to remain a master of his art. Surgery had long been a speciality and soon other specialities arose. The keystone of the medical art is the wise and experienced physician—and at the beginning of this epoch the western world was blessed with a galaxy of physicians who embodied all the experience of their time.

This was the era of eponymous diseases—conditions called after the names of those who had first described them. In 1827 Richard Bright of Guy's had described the particular type of inflammation of the kidney which bears his name; and more important, he had established a rational basis for the study of kidney diseases generally. In 1855 Thomas Addison described 'Addison's disease'—a rare and fatal condition due to disease of the suprarenal glands.[1] He had also described that type of anaemia which for nearly a century was designated 'pernicious'. For nearly a century it was pernicious but is so no more, thanks to the liver therapy of Minot and Murphy and the isolation of vitamin B_{12}. Hodgkin described the condition of lymphadenoma—a fatal enlargement of the lymphatic glands—which bears his name. Graves described in full the varied symptoms of exophthalmic goitre, and Sir William Stokes wrote learnedly on diseases of the heart. But to these men 'clinical medicine' was in fact 'clinical'; the patient was studied at the bedside. Apart from a stethoscope—wooden, monaural !—the physician carried few instruments. Urine examination was carried out very simply. Clinical thermometry did not really arrive until the second half of the nineteenth century; Wunderlich's classic work was not published until 1868. There was no detailed blood examination, and the use of the microscope in medicine only started about the forties.

These brilliant mid-nineteenth century physicians had, however, one great virtue: they followed their mistakes to the post-mortem room. The physician then carried out his own post-mortem examinations, and the correlation between physical signs—elicited by examination of the living patient—and the underlying morbid conditions was thus gradually established. One of the best known of these physicians who did important pathological work was Sir Samuel Wilks. There are many later examples of the part which experience of the post-mortem room plays in the making of a great

[1] Also known as the adrenal glands.

physician—notably that of Sir William Osler at McGill in the late seventies.

New Instruments and Methods

The year 1851 was a landmark in the practice of the general physician. In that year Helmholtz invented the ophthalmoscope, which enables the observer to study the retina of the living eye. This instrument is important for the general physician since many diseases affecting other parts of the body—as cerebral tumour and certain types of nephritis—leave characteristic traces in the retina. In 1855 the laryngoscope was invented. In 1845 Hughes Bennett and Virchow independently described cases of the condition which came to be known as leucaemia. The recognition had been made by examining the cells of the blood microscopically. From then on-wards other diseases of the blood were recognized by the same microscopic method, and a microscopic examination of the blood cells became a routine in the examination of many cases. In 1852 Vierordt devised a method for estimating the actual numbers of the cells of the blood and this led to the invention of the modern haemocytometer. Estimation of the amount of haemoglobin in the blood presented difficulties, but in 1878 Sir William Gowers devised the haemoglobinometer which, with modifications, is in use today.

All these new instruments gradually changed the attitude of the older physicians; and while clinical acumen was still valued highly, the seeds were being sown which led to co-operation between the physician and the laboratory worker, to undreamt of advances in medical ideas, and to the tendency to specialize which is in some ways to be deprecated.

The Medical Specialities

As an example of the development of a medical speciality we may take cardiology—the study of diseases of the heart and arteries. The old-time physicians were quite content to base their diagnosis and prognosis on information obtained by their five senses, assisted by their stethoscopes. But the various irregularities of the heart beat were not clearly distinguished. Early in the 1900s a practitioner in Burnley, Dr. (later Sir) James Mackenzie, devised an instrument by which he could obtain simultaneous tracings of the heart beat, the arterial pulse, and the venous pulse. With this instrument Mackenzie laid the foundation of modern cardiology. In 1901 William Einthoven in Holland invented the string galvanometer, by which minute electrical currents could be recorded. In 1904 he showed that this instrument could be used to record the changes in electric potential in the living human heart. This was the beginning of the electrocardiograph—an instrument which is indispensable in any cardiac unit. A great pioneer of this work in Britain was the late Sir Thomas Lewis, who in 1920 published his first monograph on the subject. Another widely used instrument is the mercury blood pressure apparatus; its invention by Scipione Riva-Rocci in 1896 made measurement of the blood pressure an easy bed-side procedure.

Radiology

Just before the turn of the century an entirely new diagnostic method was born. On 8th November, 1895, the professor of physics at Würzburg, Wilhelm Conrad Röntgen, discovered a new type of ray which would penetrate opaque bodies. On 28th December, 1895 he published his paper on these 'X-rays', and by that time he had already investigated many of their fundamental properties. Röntgen showed X-ray photographs of the human hand. On 7th January, 1896, Campbell Swinton, an electrical engineer, took the first X-ray photograph for clinical purposes. From then on the new procedure made rapid strides.

At first used only in the diagnosis of fractures and the localization of foreign bodies, the X-rays were soon used in the diagnosis of diseases of the lungs and urinary tract. As early as 1898 Walter Cannon used the X-rays to study the intestinal movements of the cat. He gave the cat a meal containing bismuth—an opaque substance—and the alimentary tract was thus clearly outlined. Cannon was thus the inventor of the barium meal—for barium soon replaced bismuth as the opaque substance—which is of such vast importance in the diagnosis of gastric ulcer and cancer. In 1918 Dandy devised a method of examining the ventricles of the brain by means of X-rays. In 1922 Sicard and Forrestier used lipiodol—an oil containing an iodine compound—to outline the bronchial tree, and this method is now important in the diagnosis of bronchiectasis and other lung conditions.

Allied to the diagnostic practice of radiology are the therapeutic practices of radiotherapy and radium therapy. It was early recognized that X-rays can produce persistent dermatitis and fatal blood changes. As early as 1896 these harmful properties were used in the United States, Germany and France for the treatment of cancer. Radium therapy began from an accidental burn which the French scientist Becquerel sustained through carrying radium in his waistcoat pocket. From this developed the use of radium needles for cancer, the radium bomb, and the employment of very high X-ray voltages for deep X-ray therapy.

Of the branches of medicine which have been revolutionized by X-rays, gastro-enterology is one of the most notable. Whereas not long ago the differentiation between gastric cancer and gastric ulcer was made by chemical tests, the modern application of X-rays has made the answer much more accurate.

Nutrition

In the eighteenth century James Lind had shown that scurvy can be prevented and cured by giving lime-juice. In 1881 Lunin showed that animals cannot live on a synthetic milk diet. A few years later Takaki reduced the incidence in the Japanese navy of the disease beri-beri by changing the diet of polished rice. About the turn of the century Eijkman and Grijns showed that beri-beri was due to the lack of some substance removed from the rice in polishing. Between 1906 and 1912 Sir Frederick Gowland Hopkins laid the experimental foundation of our knowledge of the accessory food factors, or vitamins, and some years later Sir Edward Mellanby began to publish his important work which showed that rickets is also a deficiency disease, due to lack of a fat-soluble vitamin such as is found in cod-liver oil. Many accessory food factors have now been isolated, and some have been synthesized not only in the laboratory but also as a commercial proposition.

Endocrinology

The ductless glands—the thyroid, parathyroids, suprarenals, etc.—had long been a problem in medicine. It was known that their overgrowth or under-activity gave certain recognized conditions, such as myxoedema, cretinism, gigantism, diabetes mellitus and Graves's disease. The first success in treatment was achieved when Murray in 1891 cured myxoedema by giving dried thyroid gland by mouth. In 1902 Bayliss and Starling, working at University College, London, showed that a substance which they named 'secretin' is present in the mucous membrane of the duodenum, and that it is carried in the blood as a chemical messenger, or *hormone*, to stimulate the pancreas to activity. It was then shown experimentally that glands such as the thyroid, parathyroids and the 'islet' portion of the pancreas exercise their effect by pouring their secretions directly into the blood stream.

This work stimulated the search for these hormones. Adrenaline had been shown to be present in the suprarenal in 1895, and it was isolated in 1901. In 1914 the thyroid hormone, thyroxine, was isolated. In 1909 Sir Henry Dale isolated an active hormone from the pituitary. The most pressing problem was obviously that of the 'islet-tissue' of the pancreas, for this controlled the onset of diabetes. Many experimenters failed, but in 1921 Banting and Best succeeded in isolating insulin, and shortly afterwards it was manufactured commercially. The discovery of other hormones followed, and some of them have been synthesized. The synthetic products sometimes have an advantage over the natural hormone in that they can be given by mouth. The hormone cortisone, which is produced by the cortex of the suprarenal gland, is important in the treatment of rheumatoid arthritis. It promises to be one of the most important therapeutic agents of recent years. (See also Chapter XVI.)

Allergic Conditions

In 1819 John Bostock of Guy's described the condition popularly known as hay fever, and in 1831 John Elliotson found that it was due to pollen. Eight years later the great French physiologist, François Magendie, showed that if egg-albumin was injected into a rabbit, a second—or at least a subsequent—injection of the same substance would cause death. This was the first experiment in anaphylaxis, the body-state caused by the injection of a foreign protein, which was later to assume great importance when injections of foreign proteins became common in sero-therapy. Asthma had for some time been suspected as being associated with some peculiar form of sensitivity, and in 1860 Salter called attention to its association with animal emanations. In 1902 Charles Richet and Portier discussed anaphylaxis fully, and five years later von Pirquet coined the word *allergy*.

The manifestations of allergy are many, and its exciting causes legion—as anyone who has had to undergo allergic skin tests will have realized. The ramifications of allergic conditions cannot be followed here, but mention must be made of their connection with histamine. This substance was fully studied by Sir Henry Dale and Sir Patrick Laidlaw in 1910. It is liberated in the body under certain conditions, such, for example, as injury to the tissues. It is also liberated when an allergen, such as a foreign protein, reacts with a cell already sensitized to it. In recent years synthetic substances which are antagonistic to histamine have been produced. These substances play an increasing role in the treatment of asthma and allergic conditions generally.

Tropical Diseases

Sir Patrick Manson is universally regarded as the 'father of tropical medicine'. In 1877 he worked out the life history of the parasite which causes filariasis. The malaria parasite was first seen in human blood by Alphonse Laveran in 1880. The Italian scientists Marchiafava and Grassi did important work on the spread of malaria, but the fact that it was spread by certain species of mosquito was not demonstrated until the work of Sir Ronald Ross in 1897. This work was based on, and inspired by, a hypothesis which had been propounded by Manson years before.

Trypanosomes had been seen in the seventies in the blood of various animals; and in 1895 Sir David Bruce showed that nagana, a disease of domestic animals in Zululand, is due to *Trypanosoma brucei*, carried by a species of tsetse fly. In 1903 Bruce showed that another species of tsetse fly spreads the human sleeping sickness of Uganda. In 1887 he had discovered the causative organism of Malta (undulant) fever. Walter Reed and his colleagues showed in 1900 that yellow fever is carried by a species of mosquito, and the active measures which were instituted by W. G. Gorgas in 1904 permitted the construction of the Panama Canal in a region which had been a death trap.

The two world wars speeded up research on tropical diseases. Quinine and its derivatives had long been used for the treatment of malaria, but both for prophylactic purposes and for treatment they had certain disadvantages. New anti-malarial drugs such as pama-quin, mepacrin, chloroquine and paludrine have been synthesized and have proved vastly superior. Newer and more economic methods of malaria control, by attacking certain species of mosquito only, were introduced. The experiments of Fairley and his colleagues in Australia on human volunteers gave improved methods of suppressing incipient attacks, and did much to ensure victory in the Pacific.

Scrub typhus, which is due to an organism intermediate in size between a bacterium and a large virus, caused a heavy casualty list during the war until it was found that two new agents, chloram-phenicol and aureomycin, were practically specific. Suramin and pentamidine were found to be very effective in sleeping sickness, and new remedies were also used for kala-azar. In bacillary dysentery it was found that sulphaguanidine was most effective.

Tuberculosis

An entirely new method of treatment of 'consumption' was introduced by George Bodington in 1840, when he advocated—and practised—treatment of the patients in 'cold, pure air'. His work bore fruit in Germany, where Hermann Brehmer established the first sanatorium in 1859. More famous was the sanatorium at Nordach, in the Black Forest, which was founded by Otto Walther in 1888. In England the King Edward VII Sanatorium at Midhurst was founded in 1903. The lineal descendant of Bodington's institu-tion in Britain was an Anti-Tuberculosis Scheme founded in Edinburgh in 1887 by Dr. (later Sir) Robert Philip; part of this was the Royal Victoria Hospital. Included in the scheme also was the Victoria Dispensary for Consumption, the forerunner of the many tuberculosis dispensaries established later. In 1915 Sir Pendrill Varrier-Jones established the institution later to become the Papworth Colony, which combines treatment of the tuberculous condition with industrial training suitable for partially incapacitated patients.

Specific treatment of tuberculosis was initiated by Robert Koch in 1890. He then announced 'tuberculin', produced from cultures of the bacillus, as a cure for the disease. Few categorical statements in medicine have taken so long to confirm or disprove. Tuberculin is certainly not a cure, but in the hands of experts it can be employed effectively in certain forms of tuberculosis.

Apart from the graduated rest of sanatorium treatment, therapy has been most effectively based on collapse of the diseased lung. Although James Carson had suggested artificial collapse of the lung in 1821, it was not until 1888 that the operation, consisting of the introduction of a gas (usually air) between the chest wall and the lung, was put on a practical basis by Forlanini. In suitable cases a more permanent collapse is effected by removal of portions of the ribs on the affected side. Introduced by Estlander in 1879 for another chest condition, this operation of thoracoplasty was first used in pulmonary tuberculosis by de Cérenville (1885). The modern operation was devised by Sauerbruch (1909), and his work led to the brilliant results achieved, among others, by Morriston Davies and Tudor Edwards.

The Triumph of Surgery

These observations lead us back to the surgical field. The German surgeons, having mastered Lister's technique, attacked the abdomen with great courage and success. By 1881 Billroth had excised the stomach, and operations on the intestinal tract soon became common. In this country Mayo Robson and Berkeley (later Lord) Moynihan were great pioneers. Kocher of Berne was a great general surgeon, and his work on the thyroid gland had permanent value. In the United States the Mayo brothers established a surgical clinic which is world-famous.

The surgery of bones and joints had long been attended with the risk of sepsis, but in Glasgow Sir William Macewen obtained brilliant results. Macewen's greatest work was probably in connection with the central nervous system. In 1879 he removed a brain tumour, and by 1893 he had reported his results in sixty-five operations on the brain and spinal cord. Sir Victor Horsley was also a pioneer in this field, and Harvey Cushing later established himself as the leading authority in neurosurgery. Among surgical specialities may be mentioned orthopaedics, in which Hugh Owen Thomas and Sir Robert Jones were pioneers. Plastic surgery—the most recent of these specialities—has achieved many triumphs in the hands of such men as Sir Harold Gillies and Sir Archibald McIndoe. Even the heart is being successfully attacked, and the work of Blalock in America has been outstanding in this field.

Some of the success of modern surgery must be attributed to the latest anaesthetics and to the care of the patient before and after operation. The anaesthetist of today has at his command a battery of methods and anaesthetics undreamt of fifty years ago. Curare—the South American arrow poison—may be mentioned. First investigated nearly a century ago, its action in causing temporary paralysis of muscles was well understood. But quite recently methods of purifying it—and even of synthesizing it—have robbed it of most of its dangers, and it is now used to secure muscular relaxation during a surgical operation, so that the concentration of the anaesthetic proper can be considerably reduced. Preparation of the patient for operation is now a highly complex matter, in which the biochemist is actively concerned. New methods of blood transfusion, found so successful during the war, have saved many lives before and after operations.

Blood Groups

This is perhaps a suitable place to mention the development of 'blood grouping' which has been a product of this century. In 1900 Landsteiner discovered that the blood of certain individuals contains complex substances which have the power of causing the corpuscles in the blood of other persons to run together in clumps. It will be obvious that, if blood containing such 'agglutinins', as they are called, is used for transfusion purposes, serious consequences are liable to develop in the person transfused. Landsteiner divided individuals into three groups on this basis, and other workers—especially Moss in the United States—soon raised the number of groups to four. In 1930 Landsteiner, then of the Rockefeller Institute in New York, received the Nobel Prize for his researches on blood-grouping. Despite all this work accidents in transfusion occurred. The whole picture was completely changed by the discovery of a further factor in human blood by Landsteiner and Wiener in 1940.

This is the 'Rh factor', and by using it a new and very satisfactory system of blood-grouping has been evolved. Blood transfusion is now an everyday procedure. Mention should also be made of the value of blood-grouping in medico-legal work, especially in questions of paternity.

Chemotherapy and Antibiotics

The whole of medical practice is influenced by the ubiquity of bacteria and viruses. Many drugs have therefore been tried to kill bacteria in the human body. Unfortunately, nearly all those which killed the parasites also harmed the individual who harboured them. This was the position when, in 1903, Paul Ehrlich began to use trypan-red, a benzidene dye, against trypanosome infections in mice. He then turned his attention to the organic compounds of arsenic. In 1905 the spirochaete which causes syphilis was discovered by Schaudinn, and Ehrlich set out to build an organic arsenic compound which, without being too toxic to the individual, would destroy the spirochaetes in the human body. After over six hundred of his compounds had been synthesized by complex chemical methods, 'salvarsan' was produced in 1909. At a stroke the treatment of syphilis was revolutionized.

The years following this discovery saw the introduction of synthetic compounds of great use in the treatment of tropical diseases. But in twenty-five years no one succeeded in synthesizing a compound which was of the slightest value in the treatment of the infectious diseases which are prevalent in temperate climates, such as those of Europe and the United States. Then in 1935 Domagk of Elberfeld showed that the red dye called prontosil was of value in the treatment of infections due to streptococci—for example, erysipelas and certain forms of suppuration. It was demonstrated that sulphanilamide is the active part of prontosil, and from these observations have been derived the series of 'sulpha-drugs', active against so many infections.

Meanwhile, equally striking results were being achieved from another angle. It is well known that in 1929 Sir Alexander Fleming found one of his culture plates, on which colonies of a certain organism were growing, contaminated by a mould, and that he investigated its destructive effect on the germs. It was the stimulus of war which led Sir Howard Florey, Dr. Chain and their co-workers to prosecute the complex investigation which led to the commercial production of penicillin. This substance is effective against some of the infections which resisted the sulphonamide drugs. Even more, it has simplified exceedingly the treatment of gonorrhoea; early syphilitic infection responds very effectively to it, and it promises to have important results in the treatment of the later manifestations.

Fleming's work on the action of moulds in inhibiting the growth of organisms has opened up vast fields. Other moulds have been actively investigated with successful results. Streptomycin, derived from *Actinomyces griseus*, has been effective in certain acute and generalized forms of tuberculosis, and hopes are raised that a similar remedy may yet be found for the more chronic forms in the lungs, bones and joints which with our present methods take so long to cure. Mention has already been made of the almost specific effect of chloramphenicol, another antibiotic, on diseases of the typhus group; and yet another, aureomycin, is being actively investigated as a remedy in certain virus infections.

The Future

In this chapter it has been possible to touch only on some of the outstanding advances of the last hundred years. During the whole of this period continuous progress was made in prevention, by improving the environment (housing, sanitation), by the control of infectious diseases, by application of the newer knowledge of nutrition, by periodic medical and dental inspections of infants and children, and by inculcation of the principles of hygiene. How can justice be done in a few words to the effects of antenatal care in reducing maternal and infant mortality? Maternal mortality today is less than a quarter of what it was fifty years ago. In the same period the infant mortality has fallen by two-thirds. The sulphonamides have made puerperal fever—once the dread scourge of hospital obstetrics—a rather rare disease against which the obstetrician now has an effective remedy.

With public health and social medicine established on a sure basis, we have reason to hope for further advances in prevention. Healthy co-operation between physician, surgeon, pathologist and biochemist will result in further advances in treatment. The era of specific remedies for specific diseases is with us, and the near future will certainly see the synthesis of new chemotherapeutic products. It is now some years since surgery was made 'safe for the patient', as Moynihan said; and the fact that the patient is now being made 'safe for surgery' will lead to further advances in this dramatic field.

CHAPTER XVIII

GENERAL PSYCHOLOGY

by Sir Cyril Burt, D.Litt., D.Sc., LL.D., F.B.A.

PSYCHOLOGY is one of the oldest departments of knowledge. Yet from the time of Aristotle down to the middle of the nineteenth century it remained, as it had begun, a branch of philosophy rather than of natural science. Then, with surprising speed, the method of approach was altered; and fifty years later psychology had come to be recognized as a science in its own right, with laboratories and professorial chairs in France, Germany, America and Great Britain. The year 1850, with which this review commences, roughly marks the turning point.

The work of the older philosophers was by no means wasted. They had done useful service in describing, classifying and naming the chief mental processes; and had then gone on to construct a working theory of the mind. Here, however, their efforts led, not indeed to any measure of agreement, but to a fairly clear formulation of what the chief alternatives might be.

The Pre-Scientific Stage

At the time our survey opens there were three main schools of thought. The oldest and most popular view, which went back to the days of Plato, depicted the mind as made up of a limited number of separate faculties. Just as the body consisted of a number of limbs and organs, so the mind was supposed to contain between twenty and thirty faculties or 'powers', about half being intellectual or 'cognitive', and half practical, 'active' or 'appetitive'. The most familiar version of the doctrine is commonly known as *phrenology*, which still had a widespread vogue even in 1850. On the intellectual side it distinguished something like fifteen faculties arranged on three main levels—first, the lowlier capacities for sensation and perception (sight, hearing, etc., sense of position, and the like);

secondly, the intermediate capacities for memory and imagination; and thirdly, the highest powers peculiar to man, such as conception, judgment, and reasoning; similarly, on the active or emotional (now more usually called the 'orectic') side, there were first, the lower appetites and animal propensities; secondly, the various sentiments and passions; and thirdly, governing all, the supreme faculty of will. Each of these faculties was supposed to have its special organ in the brain, so that by observing the size of the several organs (shown by protuberances or 'bumps' upon the skull) the special gifts and disabilities of each individual could be diagnosed.

In contrast to this popular and pluralistic view, the academic school, deriving mainly from Kant, was definitely monistic: it insisted first and foremost on the unity of the mind. Instead of separate organs for different forms of knowing, feeling, and willing, Kant had maintained that there were merely three inseparable aspects to be found in every conscious process—cognitive, affective, and conative—a kind of three in one. The fundamental function was a unique activity technically termed *apperception*—a word which broadly designates what later writers described as *attention*.

The third, and typically British, school denied the existence both of separate faculties and of any unitary entity such as the mind or soul. What was popularly called the mind was held to be simply a mass of innumerable ideas, linked together by processes of association. This view, often called *associationism*, became the dominant doctrine during the middle years of the nineteenth century, and was not unfairly described as 'a psychology without a soul'.

In spite of their apparent differences, each of these three schools had one characteristic in common. All based their conclusions almost exclusively on introspection and armchair argument. The mental philosopher was content to sit back and reflect on his own conscious processes, and treat them as typical of all mankind. He made little or no attempt to observe the behaviour of other people, much less to submit the conclusions reached to the decisive test of experiment. And, since philosophers are intellectual people, the doctrines they evolved were essentially intellectualistic.

The Experimental Approach
The idea that psychology might be converted from a branch of

S

philosophy into an experimental science starts with Fechner, who in his earlier years was Professor of Physics at Leipzig. He tells us how, lying in bed on the morning of 22nd October, 1850, he began to ponder on the relation between body and mind, and came to the conclusion that it might be summed up in a single quantitative law, which could be verified by measuring simultaneously the simplest types of mental process and the corresponding types of physical stimuli that gave rise to them. There and then he sketched out a programme of research to test his theory. The experiments occupied the greater part of the next ten years; and the results obtained show that, in every sensory field, the intensity of a sensation varies pretty closely with the logarithm of the intensity of the stimulus: that is to say, for the strength of a sensation to increase by equal amounts, the strength of the stimuli must increase by proportional amounts. Thus, if a room is lit by 100 candles, the addition of one more will make it just perceptibly brighter: if it is lit by 200, one more will make no perceptible difference; we must add two; if it is lit by 300, we must add three; and so on. His conclusions were published in his *Elemente der Psychophysik* in 1860, an event which is usually looked upon as 'marking the birth of the new science of experimental psychology'.

Nor were his researches limited to the simpler processes of sensation. He extended his experimental methods into the field of aesthetic psychology, where he also succeeded in introducing quantitative devices. The somewhat mystical doctrines about the relation between mind and matter which Fechner based on his results now have little more than a historical interest. His real achievement was to prove that mental phenomena could be measured, and to develop ingenious techniques for this purpose—the so-called *psychophysical methods*, which have formed the foundation of all experimental psychology from that time onwards.

In 1850, the very year in which Fechner's earliest researches were planned, Helmholtz at Berlin (where he later became Professor of Physics) succeeded in measuring the speed of the nervous impulse. This confirmed the tentative idea that the nerves or nerve-fibres might be conceived as operating rather like telegraph-wires, conducting currents or 'messages' at a finite measurable rate. There are proverbial phrases, like 'quick as winking', and 'swift as thought'.

But, by using new and ingenious apparatus measuring accurately to the thousandth of a second, the times required for all the main forms of mental reaction could now be expressed in numerical form. This, and Helmholtz's later work on vision and hearing, made it clear that the detailed nature of our conscious activities is largely determined by the anatomical structure of the sense-organs, of the nervous system, and above all of what is loosely called the brain.

Down to about 1860, the academic physiologists, under the lead of Flourens, had maintained, in sharp opposition to the phrenologists, that the brain, like the kidney and the liver, functioned as a whole. But in 1861, there died in a Paris hospital a patient who had been an inmate for nearly thirty years and whose sole infirmity was an inability to talk. After his death, Broca (whose name still survives as that of one of the founders of anthropology) discovered a lesion in the third left frontal convolution of the patient's brain. Here, it was claimed, was a quasi-phrenological organ, controlling the processes of speech.

A few years later, two German physiologists were successful in stimulating the brain by electrical currents; and by 1876, in a text-book on *The Functions of the Brain*, the British neurologist, Ferrier, was able to publish a fairly detailed map showing how the brain contained, not indeed separate 'organs' for separate 'faculties', but 'centres', or 'areas', corresponding to the various senses and the simpler types of muscular action, and indicating how these several centres seemed to be linked into a complicated network by 'association-fibres' running across the brain from one area to another.

The Biological Approach

Meanwhile, a new conception of the mind's place in nature had been developing under the influence of the early evolutionists. In 1855 Spencer published the first edition of his *Principles of Psychology*, a work which boldly classified psychology as a biological science, standing midway between zoology and sociology. His doctrines excited violent opposition from the older thinkers who held that the soul was a supernatural entity, but a few years later his somewhat speculative arguments were strongly reinforced by the factual evidence supplied by Darwin, whose *Origin of Species* (1859) and

Descent of Man (1871) rank among the outstanding contributions of the entire century.

These various empirical approaches were co-ordinated in a comprehensive treatise which may well be considered the first text-book on psychology as a natural science, namely, Bain's two volumes on *The Senses and Intellect* and *The Emotions and Will* (1855-9). The biological approach did much to correct the over-simplified and intellectualistic picture of the mind, which had been put forward by the older associationists and was still dominant in the writings of the physiological and medical schools. Spencer's favourite parallel was based on the analogy between the structure of society and the structure of the mind. Just as a highly developed society consists of a hierarchy of classes, working according to a division of labour, but organized into a pursuance of the welfare of the whole, so the mind consisted of a hierarchy of functions, co-ordinated to subserve certain dominant purposes. Thus, a new stress was laid on the unity of the mind and the purposive nature of all mental life; and this rapidly developed into a systematic attack on the oversimplified doctrines of the associationists and physiological materialists. The leader of the revolt was James Ward, whose influence in this country has lasted to the present day.

In 1875, Ward, a Lecturer in Philosophy at Cambridge, who had studied for a while in the physiological laboratories in Germany, applied for a grant to obtain apparatus for studying mental processes. Had the University authorities agreed, England would have been able to boast the first psychological laboratory. They refused. But Ward's familiarity with the latest physiological and experimental work enabled him to meet the associationists and the mechanists on their own ground, and to introduce a more eclectic standpoint. By the close of the nineteenth century, thanks largely to Ward's work at Cambridge and to that of Stout a few years later at Oxford, British psychology succeeded in throwing off the fetters of a purely mechanical associationism, which continued to hamper psychological thought both on the Continent and in America for another thirty years. The conflicting views of the empirical school and the philosophical schools were thus to a large extent rectified and reconciled.

In 1879 the first psychological laboratory in the world was opened by Wundt at Leipzig. Nearly all the leading psychologists of the following generation studied in Wundt's laboratory. The bulk of his work was on sensory discrimination and speed of reaction, with 'apperception' playing the chief part in mental organization; but experimental methods were also applied to the study of feelings.

A further step was taken with the extension of experiment to the processes of learning. In 1880, Ebbinghaus, who had recently obtained a doctorate with a thesis on 'The Philosophy of the Unconscious', happened, while travelling in France, to come across a copy of Fechner's *Elemente* in a secondhand book-shop; crossing to England he became deeply interested in the work of the British associationists. As a result he started a series of surprisingly fruitful experiments on the subject of memory, regarded primarily as the unconscious association of ideas.

Habit formation, the other chief form of learning, may be regarded as consisting primarily in the association of movements. Experiments on the acquisition of habits among animals were begun by Darwin, and were continued by another pioneer, Lloyd Morgan, sometimes acclaimed as the 'founder of comparative psychology'. A visit of Lloyd Morgan to Harvard gave rise to the more systematic techniques introduced by Thorndike—testing cats and rats by repeated trials in mazes and puzzle boxes. Hobhouse's experiments on animals (started at Oxford as early as 1887) remained unknown until the publication of his *Mind in Evolution* (1901); and were always overshadowed by his more important contributions as a sociologist. Nevertheless, they formed a remarkable anticipation of Köhler's later experiments on animal 'insight'; and helped to correct the defects of a physiological associationism as applied to the field of animal psychology.

The biological psychologists were as much interested in the mental development of the child as in the mental evolution of the race. Indeed, earlier writers, like Stanley Hall, held that the former was, as it were, a recapitulation of the latter. Moreover, the fact that the processes of learning could now be subjected to experimental study had a powerful influence on the progress of child psychology, especially in its application to the problems of the teacher. In this field the most important personage at the turn of the century was James Sully, who held the Chair of 'Mind and

Logic' at University College from 1892 until well into the twentieth century. His admirable text-books on general and educational psychology had a long vogue in this country, and his personal influence in high quarters had far-reaching effects.

Individual Psychology

Hitherto academic psychologists had concerned themselves with the study of mind in the abstract; the study of individual mentality had been left largely to biographers, novelists, and phrenologists, and indeed had been thought barely worthy of scientific notice. The first to place the investigation of individual differences on a scientific basis was Sir Francis Galton, a cousin of Charles Darwin. His earliest studies (*Hereditary Genius*, 1869, and *Inquiries into Human Faculty*, 1883) at once aroused widespread interest. He himself was interested chiefly in the inheritance of mental characteristics; and by inventing new experimental devices, like that of mental testing, and new statistical procedures, like that of correlation, he was able to show how these more elusive problems could be investigated with rigour and precision. Galton realized that one of the most accessible fields of work would be found in the ordinary school; and as early as 1886 Dr. Sophie Bryant, a disciple of Galton and in her day one of the most famous of headmistresses, initiated a long and fruitful line of research with her 'experiments in testing the character of school pupils'.

Sully and Galton between them helped to establish the Child Study Society—an association which included school masters, school inspectors, school medical officers, and education officials; and as early as 1893 teachers began to submit 'problem children' to Sully and his assistant lecturer in education, J. Mitchell, for examination and reports. Two years later Sully published his *Studies in Childhood*, which, amongst other things, urged the importance of "careful records of children's progress, carried out by pre-arranged tests, to get a sample of mental activity at successive ages". In 1897 he opened the first psychology laboratory in this country, to provide a "scientific training for the observers". W. H. R. Rivers was placed in charge, but left after a couple of years to start a psychological laboratory at Cambridge. Rivers was succeeded by McDougall, who combined his London post with the

Wilde Readership in Mental Philosophy at Oxford, where he, too, opened a small psychological laboratory. When Galton moved his Anthropometric Laboratory and his 'Eugenic Record Office' from Kensington to University College, the testing and investigation both of children and of adults was for a time regularly carried out at both the College laboratories, working in close co-operation.

The collaboration between the two departments led to an elaborate project for an anthropometric survey of the population, to be carried out under the auspices of the British Association for the Advancement of Science. On Galton's initiative a physical survey had been carried out in 1880, and he now proposed to include mental characteristics. McDougall became responsible for drafting a scheme of mental testing, and set his research-students to work on the calibration of suitable psychological tests. Under his guidance, J. C. Flugel, H. B. English and I commenced experiments in the laboratory and schools at Oxford; and with his help and advice, Spearman, who had recently returned from Wundt's laboratory at Leipzig, carried out his first set of tests in village schools nearby. Meanwhile, J. M. Cattell, who had also studied in Wundt's laboratory at Leipzig, and later on worked with Galton, had drawn up, with Galton's help, a scheme of physical measurements and mental tests which he applied to students of Columbia University in New York.

In 1901 Karl Pearson, Galton's most famous disciple, had been placed in charge of the new Biometric Laboratory at University College; and a little later put forward an ingenious device for determining 'index characters' from the correlations between physical traits, using measurements collected from criminals by Scotland Yard. The method was described in some detail in a paper he gave to the Oxford Philosophical Society; and it seemed to me that his procedure might be adapted for deducing 'mental factors' (as they were later called) from correlation tables obtained with mental tests. Galton had distinguished between two kinds of 'faculties' or 'factors'—innate 'general ability' and 'special aptitudes'; in France Binet was working out a scale of tests for 'general ability' (or 'intelligence', as he preferred to call it), which might be used for the detection of mentally defective children in the ordinary schools; and together they laid the foundations for the testing of intelligence.

The Development of Mental Testing

At that date, on the assumption that mental deficiency was a kind of disease, the diagnosis of such cases had been in the hands of school doctors, who relied chiefly upon head-measurements and the observation of so-called 'stigmata' and 'nerve-signs'. Surgeons claimed to cure the condition by cutting the skull to permit the brain to develop. Teachers and inspectors, however, had expressed much dissatisfaction over the doctors' attempts to distinguish dull and backward pupils. Eventually in 1913, owing largely to the influence of Galton and Sully, the London County Council decided to appoint an official psychologist to conduct individual studies and general surveys by means of standardized tests. This was the first appointment of its kind in the world; and as a result the London schools were thrown open as fields of research to senior students of psychology, who as a rule were themselves teachers.

Similar work on mental tests was developed by Thorndike and his pupils in America, and, when the First World War broke out, the methods developed by educationists for testing children were applied to something like two million recruits for the U.S. Army, with a view to making prompt and appropriate allocations of men and selecting suitable candidates as officers. It thus became clear that psychological techniques could be successfully adopted for adults as well as children; and, both during and after the War, considerable research was carried out in the field of vocational guidance and selection.

Finally, in 1919 Myers established in London the National Institute of Industrial Psychology, and this still remains the headquarters of psychological work in the commercial and industrial fields.

Meanwhile, a prolonged controversy had broken out between the Galton-Pearson school on the one hand and Spearman and his pupils on the other. Spearman believed in the almost exclusive importance of a single factor of general ability or 'intelligence'. Most educational and industrial psychologists followed Galton in accepting the existence of innate 'special abilities' as well. The controversy led to innumerable discussions and researches, and finally, thanks to a long succession of inquiries, mainly carried out by teachers in the London schools, a basic scheme of mental factors, both intellectual and temperamental, was eventually worked out, the whole forming (in McDougall's phrase) a systematic 'hierarchical structure'.

Dynamic Psychology

Equally sharp and equally fruitful controversies arose about the same period out of the work of Pavlov in Russia, Freud in Vienna, and the leaders of the Gestalt School in Berlin. The first signs of the conflict appeared in the fields of physiological and medical psychology.

In this country the outstanding contributions to physiological psychology during the earlier years of the twentieth century consisted in the remarkable investigations carried out by Sir Charles Sherrington and his fellow workers, first at Liverpool and later at Oxford. His dominant conception was expressed in a pregnant phrase—'the integrative function of the nervous system'. By a series of ingenious experiments he demonstrated that the over-simplified notions of neural association that still dominated the views of most physiological and medical writers were entirely inadequate; and he showed how certain general principles, such as neural co-operation, conflic, inhibition, reinforcement, and the like, could be investigated on the lower reflex level.

Much the same principles had been developed by Stout in his analysis of higher apperceptive processes. Stout's work in this country anticipated in many ways the views of the later continental critics of associationism—the Gestalt school as they are called. Their main contention is that the mind from the outset tends to perceive and organize its activities in integrated wholes. And McDougall, in his *Physiological Psychology* and later writings, showed how strikingly the newer views both of psychologists and neurologists agreed with each other, and explained much that had previously seemed inexplicable. The mind, he claimed, "must be regarded, not as part of a machine, but as part of an organism. Its organization is that of a hierarchy of psycho-physical systems, the wider systems on each higher level co-ordinating and controlling the narrower subsystems on the lower levels. Such an organization can be interpreted only if we frankly recognize that it subserves the conscious or unconscious attainment of certain functions, ends, or purposes."

The Work of McDougall

Imbued with the biological conceptions that had dominated the work of Spencer, Darwin, and Galton, McDougall devoted most of his later life to the study of affective and conative processes, which he conceived as essentially purposive. The innate bases of human motivation should be sought, so he held, in the primitive instincts which man shares with the other mammals—sex, fear, anger, gregariousness, and the like—a doctrine which he took over from Darwin's well-known work on *The Expression of the Emotions*. These instincts he regarded as playing in the mental world the part played in the physical world by directed 'forces'. In the case of man, however, the inherited instincts were, from the earliest years of life, subjected to profound modification by the 'psychic forces' present in the social environment.

These views were developed at length in his book on *Social Psychology* (1908), now in its twenty-fourth edition. As the best seller among all psychological writings, it has had a widespread influence. More than any other work, it has helped to overthrow the crude utilitarian philosophy of pleasure and pain as the sole sources of incentive, and has laid afresh the foundations of sociology, economics, and political theory.

Pavlov's Conditioned Reflexes

McDougall's purposive or 'hormic' psychology was vigorously combated by the more materialistic writers who aligned themselves with the Russian reflexologists and the American behaviourists. Pavlov, it is said, while studying at a theological seminary in Petrograd, chanced to dip into an English book by George Eliot's friend, G. H. Lewes, criticizing the traditional account of the soul in the light of physiological research. There and then he resolved to devote himself to an experimental study of the subject. In so doing, he retained to a great extent the associationist point of view which Lewes and his contemporaries had adopted. Pavlov's early work was largely concerned with the influence of mental conditions on involuntary or unconscious processes such as digestion. In 1900, while experimenting on dogs, he noticed that the salivary reflex could be excited, not only by the actual presence of food, but often by the mere expectation of food. He accordingly planned a series of systematic researches to discover how this 'psychic secretion' (as he called it) was produced, and to what extent the secretory response could be attached by habituation to entirely artificial conditions. Thus, every time the dog was about to be fed, a buzzer was sounded as a signal; and, 20 seconds later, meat was given. After a dozen experiences of this kind, he found that the dog would begin to salivate as soon as it heard the buzzer, whether or not food was subsequently presented. Thus, the signal—which might be a sound, a touch, a visible diagram, or even a musical phrase— became associated with a tendency to salivate, or, as Pavlov would prefer to say, the original innate reflex became 'conditioned' to a new stimulus. In America his doctrine of conditioned reflexes achieved a wide vogue, fitting in, as it did, with the efforts of Watson and his colleagues to secure a more objective base for psychology. But during the last few years even in America the mechanical type of behaviourism preached by Watson has come to seem inadequate, and has been followed by a school of 'purposive behaviourism', which in effect incorporates the more essential elements of McDougall's teaching.

British psychologists, whether philosophical (like Sir William Hamilton) or physiological (like W. B. Carpenter, G. H. Lewes, or T. H. Huxley), have always stressed the importance of unconscious mental processes, or 'unconscious cerebration', to use a phrase that was at one time popular. As developed by Stout and McDougall, the doctrine of dynamic psycho-neural systems—operating according to the same principles of co-operation, conflict, mutual reinforcement or mutual inhibition, that had been demonstrated among the simpler reflex activities—appeared to provide an intelligible explanation for a vast number of these unconscious processes, where emotion rather than reason takes control. Such phenomena are most readily observed in the case of mental disorder.

Psychology and Psychiatry

On the Continent, where the older intellectualist and associationist view still prevailed, these irrational features of normal and abnormal behaviour appeared for long almost inexplicable. During the first half of the present century, however, Freud in Vienna founded a new psycho-analytic school, embodying doctrines which he expressed in somewhat picturesque and sensational terms, and claiming that the mind was above all the scene of unconscious processes of a highly dynamic nature. In this country the medical profession for long bitterly opposed the new notions; academic psychologists, however, found it easy to incorporate the essential elements of his teaching, because so much of it seemed to be an elaboration of views which British writers had already reached. Today the situation is somewhat surprisingly reversed. The most active psychiatrists appear to have adopted—with little or no change, though with considerable controversy among themselves—the doctrines of Freud, Adler, or Jung; whereas academic psychologists have become increasingly critical.

The difference between the psychiatric and the psychological approaches turns largely on the attention paid to the social or cultural environment. The doctor is accustomed to treat his patient as an isolated entity. His approach is clinical, in the literal sense. He sees the sick man at the bedside or in the consulting room. The psychologist, on the other hand, argues that in dealing with mental problems such an approach is far too narrow. It is, as Ward once observed, like "trying to understand the behaviour of a magnetic compass without considering the field of forces in which it is placed".

The difference comes out most plainly in the long standing disputes over the causes and treatment of crime and moral deficiency. The doctor regards delinquency as a form of mental illness; the psychologist regards it (with rare exceptions) as a normal reaction to an abnormal situation. Psychological investigations appear to show that crime is due to a wide variety of causes. The ordinary delinquent is no more mentally diseased than the savage or the baby in his perambulator. He requires educational training rather than medical treatment, a change of social environment rather than a change of health. An unstable temperament may no doubt be partly constitutional, but whether or not the unstable individual drifts into habitual crime is due chiefly to the social conditions in which he lives, and moves, and has his being.

Social Psychology

The interest in what is called 'human ecology' arises largely from the fact that in this country psychology has been closely linked with anthropology. British investigators have long enjoyed exceptional opportunities for anthropological field work; and many of our earlier psychologists—Myers, Rivers, McDougall, Marett, for example—were also anthropologists, and began their careers with first-hand studies of primitive life in Africa, India, or the Melanesian archipelago. Thus, whereas Freud and his followers assume that mental life develops spontaneously by certain fixed stages through childhood up to adolescence and that these stages are the same all the world over, British psychologists have commonly assumed that they are far more dependent on local custom and tradition—on the cultural pattern of the particular group or region in which the child has been brought up. Unlike the medical writers, they attach far less importance to sex, and lay much more stress on the other human instincts, such as fear, fighting and the various social impulses.

In their view the dominant power of social conditions means that instruction in the laboratory or the mental hospital is not enough. An essential part of the training of the educational psychologist must consist in residing in a settlement or slum, and getting to know at first-hand something of the home-life and leisure activities of the children among whom he is to work. And the outcome of these views has been a growing interest in the psychological study of some groups and social classes, particularly in our own cities and rural districts.

During the Second World War psychologists played an important part in the British fighting services. Mental, educational, and vocational tests were applied on a vast scale to over a million recruits; and the results have been collated with the occupation, area, educational background, etc., of the various men and women. The results, when finally published, should be of much practical as well as theoretical value. Both during and since the War, the official recognition of psychologists has still further established the position of psychology as a distinct and independent science, with its own technical methods, its own trained experts, and its own special way of solving human problems.

Conclusion

Looking back, then, over the past hundred years, it may be said that psychology has successfully asserted its independence after a prolonged and vigorous struggle waged on two fronts. The first half of the period was occupied largely with the struggle of general psychology to free itself from the domination of the philosophers; the second half has witnessed a similar struggle on the part of individual psychology to free itself from the domination of the doctors. Both from philosophy and from medicine psychology has undoubtedly learnt much. But during the period with which we have been concerned the most lasting contributions have been made by those whose interests lay neither in medical nor in metaphysical problems, but whose approach was rather that of the experimentalist, the biologist, the educationist, and more recently the social worker.

Modern civilization is based on science. Down to about 1850 the most remarkable progress had been made in the application of science to physical or material problems. During the century that has just elapsed, it has become increasingly apparent that, if civilization is to continue its progress—and some might add, if it is to survive at all—scientific ways of thinking must now be applied, not only to inanimate nature, but also to man. The methods that have revolutionized agriculture, manufacture, medicine, and even war must now be adapted for the study of ourselves—of the mind and character of the individual, the family, the nation, and the race.

CHAPTER XIX

MEDICAL PSYCHOLOGY

by AUBREY LEWIS, M.D., F.R.C.P.

A hundred years ago *Medical Psychology* was the title which a philosopher (Hermann Lotze) gave to his systematic account of the nature and organization of mental activity. Like others of his own and the previous generation of philosophers in England and Germany, Lotze had had a medical training which biased him towards a physiological conception of mind: his book reflects this, for its first division is concerned with psycho-physical relations, its second with the "physiological mechanism of mental life"— including a notable chapter on space perception and tactile sensation—and its third with the development of mental life in health and disease. His psychology, although thus occupied with physiological concepts and with psycho-pathology, was nevertheless by no means materialistic: he valued *cognitio rei*, the metaphysician's intuitive knowledge, above *cognitio circa rem*, the scientist's knowledge of the external properties and relations of things. This famous book foreshadowed in its themes and approach the questions, answerable and unanswerable, which were to pre-occupy psychiatry, viewed as the application of psychology to medicine, in the ensuing hundred years.

Psychology was then mainly the concern of philosophers. Experimental psychology had not been born. Psychiatrists did not conceive it their business to develop theories of their own, but rather to adopt those provided by the philosophers, and occasionally to amend or expand these as their own observations of insanity and mental defect suggested. Thus Henry Maudsley, whose *Physiology and Pathology of Mind* appeared in 1867, was indebted to Bain and Herbert Spencer. Griesinger, the corresponding (though slightly earlier) figure in German psychiatry, acknowledged in the preface to his *Mental Pathology and Therapeutics* (1845) that he had taken his psychology straight from Herbart. Maudsley and Griesinger were characteristic of the psychiatrists of their generation in their struggles to escape from the dilemma of which metaphysics and physiology were the horns. When Maudsley wrote "Though very imperfect as a science, physiology is still sufficiently advanced to prove that no psychology can endure except it be based upon its investigations . . . no one pretends that physiology can for many years to come furnish the complete data of a positive mental science: all that it can at present do is to overthrow the data of a false psychology (of intro-spection)", he was voicing the hope of psychiatrists that the study of insanity, long divorced from the main current of medicine and bedevilled by theological notions—of demoniacal possession, witchcraft, moral degeneration and the like—should be pursued by the objective methods then achieving such notable results in other branches of science. With this went the hope that the relation between normal and morbid mental activity should be established, so that physiology and pathology between them might here also put clinical practice on a surer foundation and afford from the 'experiments of nature in disease', as Maudsley said, material for testing our generalizations about mental activity.

A further division of the 'objective' method upon which Maudsley relied for the advancement of psychiatry, was the study of "the development of mind, as exhibited in the animal, the barbarian and the infant".

In the ensuing century such notions were to be applied, with varying fortune, to the elucidation of mental disorder. Physiological psychology, pathology, animal and child psychology and social psychology, from which Maudsley expected so much, proved inadequate to the major task of making psychiatry a solid branch of knowledge, as rewarding to scientific inquiry as, say, neurology or infectious diseases. The harvest was nevertheless great. Moreover, fertile theories and methods of research, which the psychiatrists of the 1850s could not foresee, have been brought to bear on these most complex problems of human behaviour.

Experimental Methods

Experimental psychology seemed to afford at once the most rational and the most exact approach to psycho-pathology. In

Wundt's laboratory the young psychiatrist Emil Kraepelin was attracted by the study of simple reactions, and the measurement of their time-relations. For his purpose he chose reading, addition, and learning numbers by heart, and investigated the effect of various drugs upon the performance of these tasks in normal subjects. He also applied the association method, originated by Galton and elaborated in Wundt's laboratory, to the analysis of hunger, fatigue, and the effects of alcohol, paraldehyde, chloral hydrate, ether and other drugs. In these experiments he had a double object—to study quantitatively the effect of drugs on separate psychological functions, and to produce artificially by this means mild mental abnormalities which could then be studied under experimental conditions and thereby throw light on the nature of spontaneously occurring abnormalities. In the latter aim he had small success, but the method was in itself valuable: subsequent research has employed it, notably in producing, by means of the Central American drug mescalin, a schizophrenia-like condition which is transitory.

Kraepelin's studies threw light on pharmacological problems, and on work and fatigue, thus enriching experimental psychology. He was followed in them by W. H. R. Rivers, whose Croonian Lectures to the Royal College of Physicians in 1906 on "The influence of alcohol and other drugs on fatigue" reported findings close to those of Kraepelin and his associates. But such studies bore far less fruit in psychiatry than had been hoped. Forty-five years after Kraepelin's first experiments in Wundt's laboratory, his own text-book of psychiatry contained the melancholy admission—by his collaborator Lange—that experimental methods had not provided an answer to the pressing questions of mental illness. The disappointment thus avowed would today be less keen: as will be shown later, experimental methods have enriched clinical psychiatry in the last quarter of a century.

Wundt's influence upon the development of psychiatry in Germany was great, though indirect: Kraepelin—for so long its dominant figure—learnt in Wundt's laboratory methods of exact observation continued patiently over long stretches, and respect for statistical checks upon any impression or supposed pattern of human behaviour, which he carried over to his clinical studies and expressed in his famous text-book. Respect for statistics had been

T

expressed earlier by another great psychiatrist, the Frenchman Pinel: "*La médicine*," he wrote in 1809, "*ne peut prendre le caractère d'une vraie science que par l'application du calcul des probabilités*." It was, however, from Francis Galton that the statistical method in psychology received its impetus, eventually to be transmitted to psychiatry.

Francis Galton

Galton initiated so many methods and inquiries ultimately applied to medical psychology that it is difficult to overstress his importance, though he had no academic post, no students therefore to work out under his direction the full implication of his seminal ideas, and no need to limit himself to any field of knowledge. He was indebted to the Belgian astronomer Quetelet for the notion of applying to social and biological data the normal law of error, usually associated with the name of Gauss. Galton, who held the conviction that measurement is the hallmark of a full-grown science, took Quetelet's concept and used it to convert the frequency of occurrence of exceptional mental ability into a measure of its amount: originally applied to men of genius, the method was later turned to the assessment of incapacity, i.e. mental deficiency. His fundamental contributions to statistics, and thereby to psychology, need no further mention here.

His more specialized concern with individual differences between human beings, and the hereditary transmission of these, has had a permanent influence upon medical psychology. His general account of the researches he made in this field was contained in the *Inquiries into Human Faculty and its Development*. He devised mental tests whereby the qualities of a population could be measured; he inquired into the forms and idiosyncrasies of imagery; he made an experimental study of association. All of these have had much to do with later investigations into psycho-pathology: the last of them calls for fuller examination, since it represents one of his ideas that proved most fertile to psychiatry.

The philosopher-psychologists, during the heyday of associationism, managed to develop their theories without any experiments, even though some advocated, as John Stuart Mill did, that the laws of this 'mental chemistry' must be found by direct experiment.

Galton went another way to work. He prepared a list of seventy-five words, each written on a separate slip of paper. Looking at each word in turn, and holding a small chronograph in his hand, he recorded the length of time it took him to be aware of a couple of ideas arising in his mind in direct association with the stimulus word. In tracing these associations (on which he made introspective observations also) he discerned "how whole strata of mental operations that have lapsed out of ordinary consciousness admit of being dragged into light, recorded and treated statistically, and how the obscurity that attends the initial steps of our thoughts can thus be pierced and dissipated. I then showed measurably the rate at which associations sprung up, their character, the date of their first formation, their tendency to recurrence, and their relative precedence. . . . Perhaps the strongest of the impressions left by these experiments regards the multifariousness of the work done by the mind in a state of half-unconsciousness, and the valid reason they afford for believing in the existence of still deeper strata of mental operations, sunk wholly below the level of consciousness, which may account for such mental phenomena as cannot otherwise be explained." This is a pregnant sentence, in the light of what 'free association' was to yield in the hands of the psycho-analysts.

Galton's experiments were promptly taken up in Wundt's laboratory, and a series of studies published by his associates. Some of these who were psychiatrists, observed the content and time-relations in experiments with free and controlled word association in various mental disorders, such as catatonia and mania, and by this means endeavoured to make diagnosis less a matter of clinical art. Further work on these lines was carried out in the United States, notably by Kent and Rosanoff: the method had been developed and introduced into American psychology by Cattell, a pupil of Wundt, who had been in personal contact also (while lecturing at Cambridge in 1888) with Galton.

Word Association Studies: Jung

The method was given a new purpose by Jung and Riklin, the Zurich psychiatrists who in 1904 published their studies in word association. These reverted in part to the classical associationism, but made a bold departure by assuming that deviations from normal response—delay, complete silence, repetition of the stimulus-word, misunderstanding, inappropriate laughter, rhyming and so forth—were due to a conflict of which the patient was often unaware but which the stimulus-word somehow touched and activated. Jung used the word *complex* to denote the system of desires, emotions and memories thus stirred up. The psychogalvanic reaction was found to be correspondingly disturbed when a complex was touched, with consequent emotional upset. Jung's findings have been modified by subsequent workers, but their importance lay in reviving and clarifying Galton's emphasis on the unconscious determinants, and in using the method for the study of mental disorder. Jung found that the responses indicating a complex occurred more frequently among mentally ill persons than among the healthy, and that there were in schizophrenic patients more responses dealing with inner feelings, whereas manic-depressive patients gave an unduly high proportion of responses concerned with external objects. From these findings he arrived at his formulation of the psychology of dementia praecox and the types of personality which he called *introvert* and *extravert*. Jung's work and theory seemed at that time to bring him close to Freud's psycho-analytic interpretation of morbid and normal mental activity, then comparatively little known. The subsequent development of psycho-analysis has made it the most potent overt influence in medical psychology during the last thirty years.

Freud and Psycho-analysis

It is truer of psycho-analysis than of any other major psychological system that it is one man's work. Freud had gifted and industrious pupils who enlarged—and in some cases reframed or undermined—the remarkable structure he put up: but in design and in execution it was predominantly his work. He was subject to many influences—from his teachers Brücke, Meynert, Charcot; from his associates such as Breuer; and from others less easily identified; and his theory is indebted to the scientific and cultural climate in which he grew up. But the distinctive bias of his fertile and courageous mind is impressed on the curiously inseparable conglomerate of clinical treatment, metaphor, dynamic theory and technique of investigation which is all comprehended within the term 'psycho-analysis'.

The stress laid on 'the unconscious' in psycho-analysis was far greater than in any other psychological system, and the qualities and powers attributed to it were without parallel in earlier writings: but it is to put this cardinal notion of psycho-analysis out of perspective to forget that Herbart had concerned himself with the nature and fate of ideas repressed from consciousness, and the dynamic factors responsible for the 'opposition' between ideas which, as he supposed, makes some become unconscious. The Danish psychologist Höffding in 1887 published his *Outline*, in which he stressed that 'subconscious' processes of which we are unaware play a determining part in our conscious activity.

Freud, using free association as his key, opened doors upon unconscious mental happenings which led him to emphasize the role of sexuality in many forms during the stages of individual human development. It is superfluous to recount the now familiar views which he elaborated and put forward. During most of his lifetime there was no substantial departure—within the fold of psycho-analysis—from the conceptual system he originated: those who deviated materially from it, seceded and founded schools of their own. But now the modifications or additions officially taught are considerable, and certain of these appear to some to be in part fundamentally opposed to Freud's teaching.

Freud was primarily an observer who was less concerned about consistency in his theory than about the interpretative value of his concepts for the phenomena he found in the dreams, symptoms and free associations of those he analysed, or in the comparable material available in literature whether as case records, autobiography or folklore. The provisional form in which he cast much of his exposition and the absence of sharp definition in his terms accounts for the difficulty there has been from the beginning in subjecting his conclusions—and, of course, those of his followers—to scrutiny by the methods commonly employed in scientific validation. It is beyond dispute that the exact methods of experimental psychology had failed to illuminate the most important problems of human personality and behaviour—passions, emotions, aberrations, the springs of conduct—and that psycho-analysis has opened up this region with most powerful and stimulating effects upon the study of psycho-pathology and other social and psychological subjects. So vast an undertaking was achieved in one man's lifetime by methods, undoubtedly painstaking and brilliant, but variable and highly personal. The result has been a situation for which parallels can more easily be found in religion than in science.

So long as psycho-analytic theory was mainly the outcome of treating people with neurotic illness by a new method, its merits were judged—apart from irrational opposition—by applying medical criteria: did the patients get better? could their symptoms be fully accounted for by the new explanation? did it conform to other biological knowledge and methodology? As the theory developed, and fresh territory was explored in child development, delinquency, religion, anthropology and other non-medical directions, more penetrating critical methods for appraising its value had to be looked for.

The growth of psychology had provided objective techniques, unknown to the earlier workers in experimental psychology: it was reasonable to hope that these could be applied to the verification of Freudian tenets. This was necessary, since, in the eyes of the experimentalist and the natural scientist generally, psycho-analysis has relied upon techniques that are scientifically bad—that "do not admit of the repetition of observation, that have no self-evident or denotative validity, and that are tinctured to an unknown degree with the observer's own suggestions".

Experimental Evidence

In attempting to check psycho-analytic findings by experiment, the investigator is hampered in three serious respects—the psycho-analyst's language is not sufficiently precise, nor his definitions 'operational' enough, to enable his assertions and conclusions to be clearly verified or disproved (except perhaps by other psycho-analysts); the nature of his facts, like those of other physical, biological and social scientists, is in part a function of the method by which they are obtained (i.e. by psycho-analytic procedures, notably free association and interpretative observation of free activity in children). While psycho-analysts are agreed on many broad generalizations, on details and on some fundamental points of theory they differ greatly, leaving the non-analyst in doubt whether they can in fact all be using the same technique of investigation and interpretation, and in doubt also which of the disputed theoretical statements he can take as the proper subject of his experimental or verifying study. Hence the extent to which psycho-analysis has been confirmed or falsified by extraneous studies is, as yet, disappointingly small. Sears, who has brought together much of the relevant material, is forced to the conclusion that, in spite of the volume of experimental study of such matters as frustration, aggression, displacement, repression and projection, little more has been achieved than to give crude confirmation of some psycho-analytic findings generally accepted, but it has not been possible to verify the detailed assertions about these and other mental happenings and forces which are the substance of psycho-analytic theory: "it seems doubtful whether the sheer testing of psycho-analytic theory is an appropriate task for experimental psychology."

The invigorating effect of psycho-analysis upon medical psychology and on psychology in general has been so great, and the doubt about many of its tenets and about its efficacy as a therapeutic agent so persistent during the last fifty years, that—in spite of growing acceptance—there is need for purification of the psycho-analytic method of investigation (and treatment) so that those who use it will more readily agree, and be explicit and strict in their formulation of the theory, thereby enabling its application where non-psycho-analytic tests of its predictive correctness can be employed on a sufficient scale. This is an exacting demand. It cannot be met by animal studies, which are only to a limited extent appropriate to a theory that makes human consciousness and unconsciousness a cardinal feature of its conceptual structure: on the other hand the complexity of human mental activity may make it impossible to test explanations that are ambitiously concerned with the whole genesis of men's motives and behaviour—impossible, that is, unless the subjects can be observed over a period of years and will permit close examination of themselves and their lives. It is significant that Freud, in his latest exposition of psycho-analysis, wrote "The teachings of psycho-analysis are based upon an incalculable number of observations and experiences, and no one who has not repeated those observations upon himself or upon others is in a position to arrive at an independent judgment of it." It is the independence, however, of the judgment formed by the initiate that is often in question.

The impetus given to medical psychology by psycho-analysis is so indisputable and any reluctance by psychiatrists to accept the teachings of psycho-analysis apparently so ungrateful that it is proper to stress the distinction between serviceable techniques and stimulating hypotheses on the one hand, and a psychological science on the other. If psycho-analysis had been more scientific, it would not have opened up so much unknown territory in so short a time: but neither would it have remained so elusive, and protean to the inquirer from without:

hic tibi, nate, prius vinclis capiundus, ut omnem
expediat morbi causam, eventusque secundet . . .
tum variae eludent species atque ora ferarum.

Variants and Offshoots of Psycho-analysis

Of those who were once psycho-analysts, the most prominent and the one who has travelled furthest on his own road is C. G. Jung. His 'analytical psychology' is now remote from Freud's teachings, though not so irreconcilable with the newer theories put forward by Melanie Klein. Important and therapeutically helpful though these are held to be by many who by temperament and the nature of their personal experience of psycho-analysis are favourable to them, they are by their nature and origins even less susceptible of independent verification than Freud's theories, and it would be a fantastic extension of the term to class them as scientific. Their kinship is rather with philosophy, folklore, mysticism, and the imaginative arts.

The many offshoots and variants upon the psycho-analytic theme—commonly described by the name of their chief proponent, such as Adler, Stekel, Rank, Horney, Alexander and, as already mentioned, Melanie Klein—have, like their parent, zealous and convinced adherents, and many striking therapeutic successes to their credit: their very number and self-sufficiency, however, are an index to the weakness of their claim to be scientific, in the usual sense of the word.

The French Contribution

Contemporary with Freud and like him influenced by Charcot, the French psycho-pathologist Janet elaborated a descriptive and analytic study of neuroses which put much emphasis upon the part played by psychic tension in effecting healthy synthesis in mental functions. His concept 'dissociation', invoked to account for restriction or apparent fragmentation of consciousness—such as hysterical patients exhibit with their 'loss of memory' and 'multiple personality'—attested the impress left by Charcot's choice of clinical interests upon his pupils; much of Janet's work also bore witness to the teachings of Th. Ribot, who from 1889 held a Chair of experimental and comparative psychology at the College de France and who had introduced to France English psychology, as well as, later, the work of Wundt and his forerunners. Taine, Ribot and Janet were all convinced that study of the abnormal was one of the most fertile methods of elucidating general psychology. Binet, too, whose development of intelligence tests has made his name famous, had at first a similar interest in psycho-pathology and hypnotic experiments, which gradually passed into an absorption in the problems of personality, of stupidity and genius, and finally the experimental study of intelligence. It is only this last great outcome of French interest in medical psychology which—like the earlier French contribution to the understanding and education of the mentally defective which Edouard Seguin carried to the United States—has had an evident effect upon the practice and scientific development of psychiatry in the Anglo-Saxon or in the other Latin countries: for the most part Janet's impressive contribution to psycho-pathology and psycho-therapy has been overshadowed outside France by psycho-analysis and its offshoots.

Typology

The effort to classify people according to their temperament, which has such a long history, has from the outset been related to mental disorder: such ancient words as melancholia attest this. Psychiatrists and psychologists at the end of the last century were ridding themselves of some of the cruder typological notions, e.g. those of Lombroso, and the way was prepared for a re-classification of people and diseases on better lines. Kretschmer, a psychiatrist, related the manic-depressive psychosis and dementia praecox, the two great psychoses classified by Kraepelin, to a particular mental constitution and to a physical structure or habitus which he described. Jaensch classified people according to their imagery. Italian writers investigated actively certain mental and physical correlations also for typological purposes. Recently Sheldon has made large and more exact studies of measured constitution. As statistical methods appropriate to the task have been elaborated, much productive effort has been put into the classification of traits and other taxonomic work designed to improve the working categories which medical psychology uses.

Animal Psychology

Animal psychology has yielded some information that is relevant to psychiatry, and influenced psychiatric theory. Much of this is concerned with behaviour that can be equated descriptively with neuroses, for scientific purposes in which the subjective features of the latter are ignored. Rats, for example, when subjected to irritating stimuli and unable to escape from them, showed a sudden burst of violent undirected running, which might go on until they had a convulsion. Cats who were regularly subjected to a blast of air that frightened them when they went to take food manifested restlessness, trembling, hiding, marked startle response to minor sensory stimuli, and disturbances of pulse and respiration; they showed anxiety and tried to escape when they were given the signal for feeding, they refused to eat, and in many cases courted an unusual amount of fondling by the experimenter but reacted aggressively towards other animals, and they followed certain 'compulsive' rituals: their disturbed behaviour could be allayed or abolished by procedures akin in some basic respects to those of psychiatrists treating human neuroses. Many studies of this kind have now been carried out with different animals.

Pavlov was the first to pursue them explicitly: his observations remain significant as well as historically important, but his theoretical concepts, as of internal inhibition, extinction, reciprocal induction, and excitation have lost ground. It has become manifest that the responses called neurotic can be evoked by frustrating situations; they are unadaptive; they vary from animal to animal sufficiently to indicate that the properties of the individual determine the selection of specific behaviour more than the situation does; they are compulsive; and they are eventually exhibited also outside the experimental situation. The most recent writer on the subject, who is also one of the most experienced scientific investigators in this field (N. R. F, Maier), holds that such studies disprove the thesis that neurotic behaviour, like other behaviour, is "problem-solving and goal-oriented in nature"; neuroses can better be elucidated by frustration studies than by search for motives: "life histories contain so much material that bias invariably functions in determining which data are pertinent and which are irrelevant. Thus a theory can often be read into a case. Experimental evidence and research must establish the guiding mental sets for the psychiatrist and determine what factors in life histories are pertinent. Animal experimentation may be a crucial type of research, since it is less hampered by mentalistic concepts and introspective reports, which are often vague and spotted with rationalizations."

Behaviourism, which was developed by Watson as a psychology without reliance upon subjective material drawn from consciousness, a psychology of stimulus and response, has much influenced "common-sense psychiatry". In the hands of later investigators like Lashley it has cast off some of its dogmatic features and become part of experimental psychology, with a vigorous repudiation of psycho-physical dualism in abnormal as much as in normal psychology.

Gestalt Psychology

Gestalt psychology provided insight into the phenomena of perception, learning and thinking, which has been applied, with much profit, in the study of perceptual and other disturbances in cerebral disease or trauma, and in schizophrenia. It has latterly, through Kurt Lewin and his pupils, thrown light on dynamic problems which are important in psychiatry, stressing always the environmental situation or behavioural field, the 'valence' of a region within the 'life-space' of an individual which attracts or repels him, whereas within the individual there is a tension characteristic for each activity. This 'topological psychology' which turns its face away from the genetic or historical aspect, and therefore from the traditional medical view of human behaviour, has been studied experimentally, giving rise to measures of the level of aspiration, effects of interruption of activity, and other matters important for the study of personality and motivation, especially in social relations.

It is too early to judge the value of its applications to psychiatry, but its promise is considerable.

The Biological and Sociological Background

There is a biological as well as a mathematical background to the conceptions of Lewin. A biological frame of reference is also plain in psycho-analysis, in the ecological study of mental disease, in many comparative studies through animal experiment, and in genetic studies of constitution in relation to mental abnormality. The biology is sometimes as implicit as the metaphysics; implicit or explicit, it is often outmoded and has not been radically modified to conform to the later developments of biology as well as of psychology and medicine. Nevertheless the influence at various stages of Darwin, Lloyd Morgan, Weismann, Loeb, Jennings and many more, has determined not only the direction and conceptual framework of investigations by men like Thorndike, but has also been responsible for current psychiatric theories, often loosely grouped as 'psycho-biological'. Adolf Meyer and Henry Maudsley, each in his generation, illustrate this.

Social psychology, of which the British exponents have been prominent (Rivers, C. S. Myers, William McDougall, F. C. Bartlett) has ceased to be a largely independent branch: it has established close links, of method and subject matter, with anthropology, psychometrics, animal psychology, psycho-analysis, Lewin's vector psychology, and sociology, so that its boundaries are now ill-defined and its application to medicine rather more peripheral than is good for psychiatry—or perhaps for social psychology: psychiatry is pre-eminently a matter of social relations, and illustrates many aspects of their breakdown, as well as of their comparative stability under stresses.

Mental Tests

Child psychiatry has been heavily indebted to genetic psychology and 'mental tests' for the same reasons, and by the same routes, as has educational psychology. The mentally defective child, the maladjusted child, the child predisposed to mental illness (whether this be classed as neurotic or psychotic) can only be fully observed, its functional capacities measured, and its difficulties relieved or prevented, by the application of such knowledge as has been made available by the observational methods of investigators like Preyer, Wm. Stern, Arnold Gesell, Bühler, Piaget, and Susan Isaacs, who observed, described and measured development by various means. Psycho-analytic and kindred theories and practice have latterly prevailed in a majority of psychiatric clinics for children so far as interpretation and individual treatment are concerned, but there is more and more dependence on the use of tests, for assessing not only intelligence in its various aspects but also the structure and trends of personality and therefore the needs of the child and to some extent its probable future.

The development of mental tests, of which Galton and Binet may be considered the parents, has been so rapid and vast that application of many of them for psychiatric purposes has been at times uncritical, and at times tardy, or overhasty. Some of the methods, conversely, are of psychiatric provenance: the Rorschach inkblot, for example. The proper use of a test in psychiatry is inseparable from understanding of the psychological theory and of the statistical techniques upon which the particular test is based; it is also inseparable from understanding of the clinical applicability of the test in question. These requirements are not always easy to satisfy: but when they are, the clinical value of tests and their great scope as means of research in psychiatry has been demonstrated beyond dispute.

Statistical Methods

Statistical methods, indispensable in the design and validation of tests, are now also indispensable in studying mental structure and the patterns of disease. To a large extent the methods are those employed in all biological research, and attest once more the influence of Galton and such successors as Karl Pearson, Udny Yule and R. A. Fisher: but they have also been adapted for peculiarly psychological purposes. The methods thus developed by Godfrey Thomson, Spearman, Burt, Thurstone, Guilford and their associates have been turned to psychiatric problems closely interwoven with the texture of personality and the emotional variations that occur in human beings. Eysenck and his co-workers in London have been particularly active in this way. It is too soon to judge how far-reaching will be the influence of such statistical psychological methods upon psychiatric notions of constitution, objective measurement, and psycho-physical relations: there is good reason to believe it will introduce simpler but less vague concepts and demonstrate more manageable relations than those in current use.

Statistical research in psychiatry, along these lines, is associated with the employment of scientific methods of observation, such as those of physiology, anthropometry, biochemistry and sociology, as well as the strictly psychological methods. It is in the line of development which psychological and social studies, and to some extent psychiatric studies, pursued with vigour during the late War. The impetus has not been lost.

In this brief account, the growth of psychiatry through studies in neuropathology and neurophysiology, biochemistry, heredity and pharmacology (as well as through clinical intuition and therapeutic empiricism) has not been mentioned. Only the contribution of psychology to psychiatry—or what may be called (by analogy with educational psychology) medical psychology—has been the theme. But just as it has become increasingly difficult to separate sharply one kind of psychology from another, or to discern schools in psychology rather than convergent and maturing scientific disciplines with a common subject matter, so it has become misleading to distinguish the applications of psychology wholly from those of physiology and other biological sciences when reviewing the scientific progress of psychiatry, or of medical psychology, in the last hundred years. Dualism is becoming less entrenched: the recent advances in the physiology of cerebral activity and in the physiological concomitants of emotion have here much reinforced the clinical inclinations of the psychiatrist.

The philosophic and other conceptual difficulties remain, but the investigator finds more that is common to both the physiologist's and the psychologist's sight of the organism, in health and disease, than could have been hoped even twenty years ago. With that notable advance in our knowledge of the human organism psychiatry too has moved forward, increasing its heavy debt to the basic sciences.

CHAPTER XX

THE SIGNIFICANCE OF SCIENCE

by H. Dingle

In the preceding chapters the development of science during the past century has been traced along various lines, most of which were laid down in earlier times. It has been shown how some of them have coalesced while new directions of inquiry, unthought of before, have been opened up. The change everywhere has been enormous and its consequences unforeseeable, and although, as was indicated in the Preface, a complete account is out of the question, it is impossible to read the story told here without realizing that we are in the presence of one of the most momentous of all human activities, one which seems likely to play the chief part in determining the future course of life on this planet. It would therefore be unfitting to close the account without making some attempt to estimate the meaning of the whole practice of scientific investigation and to understand as far as may be what we are doing when we engage in this rapidly accelerating progress towards an unknown goal.

The Philosophy of Science

This inquiry, this process of self-examination so to speak, on the part of science, which has come to be known as *the philosophy of science*, may indeed itself be treated as a developing thing, and no more suitable starting-point for tracing its history could be chosen than the period that saw the Great Exhibition of 1851. For the first major work in which the philosophy of science was treated as a separate subject, distinct from general philosophy on one hand and the actual practice of science itself on the other—namely, Whewell's *The Philosophy of the Inductive Sciences*—appeared in 1840. Up to that time—and indeed, in the minds of most thinkers for many years after—the question, What are we doing when we engage in the pursuit of science? was ignored not so much because it was regarded as unimportant as because the answer was thought to be obvious. It is by no means obvious to us now and our own generation has witnessed many specific attempts to answer it, but when we look for the corresponding attempts a hundred years ago we find them almost non-existent. To understand what the men of the mid-nineteenth century understood by science we must depend for the most part on the implications of their casual remarks, their asides, thrown off in odd moments of reflection and comment on the remarkable things that they were bringing to light. And for these in their turn we must look not so much to the recognized vehicles of communication of scientific men, the technical journals and proceedings of learned societies, as to the more popular utterances in which men of acknowledged scientific standing made known the course of discovery to the general interested public.

The media of such communication were far more restricted in 1851 than they are today. Broadcasting of course was undreamed of. There was no regular scientific journal—*Nature*, which today performs an indispensable function in disseminating scientific knowledge, first appeared at the end of 1869 and for many years ran at a loss—and books on scientific subjects were both far fewer and far less accessible than now. Newspapers, it is true, gave fuller and more authoritative accounts than their average modern counterparts but their circulation was much smaller and only the more outstanding advances were reported. Apart from these, the public depended for its knowledge of scientific progress on occasional public lectures at local Philosophical Societies and Mechanics Institutes, and above all on the annual meetings of the British Association, at which, notably in the Presidential Address but also in some of the numerous other communications on all branches of science, the leaders of thought spoke to the educated layman in terms which he could understand and appreciate. It is in such sources that we can best discern the implicit view of the nature of scientific endeavour that was shared by pioneer, follower and spectator alike.

The Nineteenth-Century View of Science
The fundamental assumption, universally held as an intuitively obvious truth, was the Cartesian duality of thinking things and

extended things—mind and matter. The scientist was, in fact, a mind, and the object of his inquiry was matter, dead or living. The situation was beautifully clear-cut and precise. On one side lay the field of inquiry, the material world, or nature, standing off from the inquirer and preserving an essential and unalterable 'reality' independent of the thoughts and wishes, hopes and fears, goodness or badness of the minds that contemplated it. On the other side were those minds, whose scientific function was solely to discover, first by observation and experiment and then by rational deduction, what the material world contained and how its course was ordered. Any other activity of the mind, that could not be traced to its origin in the material world, was illusion, fit only to be ignored by the man of science.

This attitude was not confined to the 'materialist', the atheist or the sceptic. It was shared equally by the devout religious worshipper, by the artist and even by the spiritualist. To say that God was real was to say that He was a part of the external world of nature—the supreme part, no doubt, but still an object accessible to scientific discovery. To say that something was beautiful was to say that it possessed an objective ingredient called 'beauty', for which scientific tests were possible if we only knew how to apply them. To say that spirits existed apart from the body was to say that they were constituents of the natural world, located in space, and susceptible to observation by the appropriate means.

This view of things appeared so evident that it is difficult to find a passage in the writings of the time that even openly implies it; the implication lay so deep that detailed analysis is necessary to show how inevitably the whole scheme of thought rested on it. But consider the following remark from the Presidential Address by Sir John Herschel to the British Association in 1845, an Address in some ways remarkably in advance of its time, to which we shall refer again. "One of my predecessors in this Chair," he says, "has well remarked, that a man may as well keep a register of his dreams as of the weather, or any other set of daily phenomena, if the spirit of grouping, combining, and eliciting results be absent." It is here taken for granted that to keep a record of dreams is an utterly idle occupation—indeed a prototype of such occupations—and it would never have occurred to any of the audience to question the assump-

U

tion. Something like one-third of our possible experience was thus removed from scientific inquiry without even a suspicion that anything was being left out, simply because of the strength of this fundamental and unrealized Cartesian assumption.

The Law of Causality

Scarcely less deep-rooted was the belief in the inevitability of the Newtonian method of describing the course of natural events. This rested on the doctrine that the external universe consisted on one hand of inert matter and on the other of forces which moved that matter in accordance with laws which it was possible to discover. The success of this view in astronomy and mechanics was indeed amazing, and more and more phenomena were at this time appearing to be at bottom mechanical in character. In 1846 Newton's law of gravitation had its most spectacular triumph in the discovery of the planet Neptune through mathematical investigation of minute aberrations of Uranus, and it seemed as though the ultimate and universal key to the whole system of nature had been discovered. "Newton," exclaimed the Rev. T. R. Robinson in addressing the British Association as its President in 1849, "stands like a Sun in the Heaven; all is luminous after he has risen, all before darkness or twilight." The pattern of scientific inquiry was to observe the motions of bodies and to deduce the forces that made them what they were. The bodies themselves were absolutely inert, so that the relations of the forces to the motions was one of complete cause and effect. We observed the effects and our problem was to infer the causes.

This type of explanation was extended from motions to other phenomena not obviously concerned with motion at all, and the idea of forces moving bodies was generalized to that of laws of nature governing the objects of nature. For example, the temperatures of bodies were deducible from laws of heat, and those laws were accordingly granted a status equivalent to that of the law of gravitation. In this case, indeed, heat was recognized to be but motion in disguise and there was a large measure of belief in the idea that ultimately all laws were mechanical and could be reduced to terms of Newtonian forces. But while this was only a widely held probability, much less convincing in the biological than in the physical sciences, the more general view that nature was completely governed by laws was universally held. The analogy with political laws was very close and nature was looked upon as obeying the laws of nature in much the same way as a good citizen obeyed the laws of his country; it was Duty, according to Wordsworth, that preserved the stars from wrong.

Nature of Scientific Law

The inference from law to law-giver was commonly made, and the scientific writings of the time abound in references to God as the supreme legislator. "Like our soldiers and our sailors," the Marquis of Northampton told the British Association in 1848, "like the ministers of the laws of the land and the expounders of the laws of morality and religion, the inquirers into these other laws which regulate His creation . . . have duties to perform which it is criminal in them to neglect." But the political view of natural law was not confined to those who found it conformable to their religious beliefs. It was held universally, and Huxley and the Bishop of Oxford could at least agree on this, however they might differ with respect to what the laws were. "I take it," wrote Huxley in 1854, in that famous address *On the Educational Value of the Natural History Sciences* in which he describes science as 'organized common-sense'—a phrase that has been widely misapplied through ignorance of its context—"I take it that all will admit there is definite Government of this universe—that its pleasures and pains are not scattered at random, but are distributed in accordance with orderly and fixed laws." And it is well known how later he posed the problem of the conflict between the moral law and the cosmic law, both laws of the same type, namely, commands which those subject to them were called upon to obey. Herbert Spencer also, in his *First Principles* (first published in 1862) describes science as " a continuous disclosure of the established order of the universe".

The chief, if not the only, difference between scientific and political laws was that whereas human beings sometimes broke the law of the land, the world of external nature, on the other hand, was infallible; hence our power of inferring the laws from its behaviour. The connection of cause and effect was inviolable; observe the effect truly and when you have enough examples the law will be deducible as the unique condition of their possibility. It was not a mere matter of invariable sequence. The cause was a positive, existent thing, an agent in the natural world, which was not immediately evident to the senses as were its effects but was none the less an integral part of nature discoverable by reason. Nature was thus dual; there was the visible, audible, tangible course of effects apprehensible by the senses, and the remote, unobservable but equally real set of causes apprehensible by the intellect. Experience and reason were two means of attaining knowledge, and each was adapted to the discovery of its own part of nature.

Isolated Criticisms

There were indeed isolated voices raised in protest against this view. Some men of science—they can be called either old-fashioned or in advance of their time with equal propriety—held that science should discover, in Aristotelian terms, proximate and not merely final causes, that that alone should be called a 'cause' that was immediately responsible for the effect in question, and not the general 'law' of which the operation of that cause was merely one example. "I may be allowed to suggest," said Sir John Herschel in the Address already referred to, "that it is at least high time that philosophers, both physical and others, should come to some nearer agreement than appears to prevail as to the meaning they intend to convey in speaking of causes and causation. . . . The idea of *law* is brought so prominently forward as not merely to throw into the background that of *cause*, but almost to thrust it out of view altogether; and if not to assume something approaching to the character of direct agency, at least to place itself in the position of a substitute for what mankind in general understand by *explanation*. . . . A *law* may be a *rule* of action, but it is not *action*. The Great First Agent may lay down a rule of action for himself, and that rule may become known to man by observation of its uniformity: but constituted as our minds are, and having that conscious knowledge of causation, which is forced upon us by the reality of the distinction between *intending* a thing, and *doing* it, we can never substitute the *Rule* for the *Act*." Nevertheless, Herschel himself would doubtless have admitted that between his 'cause' and its immediate effect there was not merely a particular time order but some objectively existing principle of necessity, and it is difficult to see how he could have done other than identify this principle with the general law of nature that was being exemplified.

We might proceed further with the implicit view of the character of science that was held a hundred years ago, but to do so we would have to forgo universality to some extent and enter a region where dissentient voices would become more than negligible. Enough has been said in any case to give us a sufficient picture of the philosophy of science of the time and to enable us to contrast it with that required by the science of today. We see a world of external nature, sharply distinguished from subjective illusions, whose course, observable by the senses, is the result of the operation on inert matter of universal unalterable laws deducible by reason. Of that picture not a single feature remains. Our view of science is completely transformed, and although this is far from being generally realized, since the majority of scientists are not philosophically minded and are little more aware than were their great-grandfathers of the pre-suppositions of their work, the whole practice and outlook of modern science is utterly incompatible with the world-view of Victorian times. It is, in fact, much nearer to the seventeenth century views of Galileo and Newton. Let us try to trace the course of the development.

Breakdown of the Nineteenth-Century Assumptions

It did not come about as the result of any revolutionary discovery or catastrophic change in scientific practice; it was the natural and inevitable development of science itself that broke the philosophic shell in which science was enclosed. A philosophy of science should specify the basic assumptions or postulates which scientists, consciously or unconsciously, express in their practice, and this the nineteenth-century philosophy did not do. Up to the close of the century, however, the incompatibility (to change the metaphor) lay too far beneath the surface to be observed, but as the scientific probings went deeper and deeper it was inevitably brought to light and it then became clear that the apparent agreement between the fundamental principles of scientific inquiry and the world-picture we have outlined was, in fact, accidental, and it vanished completely when the later achievements of science were taken into account.

The Origin of Modern Science

We can now see fairly well how the misunderstanding arose. The lines on which science has proceeded since the early seventeenth century were laid down by Galileo and his successors, and their distinguishing feature was that the ultimate data for scientific study are our experiences—experiences of motion in the first instance, but later those of heat, of colour, of noise, and so on. It is these experiences that form the subject-matter of the various sciences of mechanics, calorimetry, optics, acoustics and the rest, and they are studied without reference to the particular material bodies that happen to exhibit them. Thus, for example, whereas in pre-Galilean science a terrestrial body was said to move differently from a celestial one, and a heavy body to fall more quickly than a light one, Galileo and Newton sought for and found laws of motion in itself, which were true whatever the location or weight or colour or shape or anything else of the moving body. The subject of study, in brief, was the aggregate of our experiences, and not any external world, or nature, which may be conceived to have caused those experiences.

The Compromise with Cartesianism

This was a new idea, so much at variance with the traditional, common-sense view of natural philosophy that it would in any case have been hard to become familiar with. That hardness became practical impossibility when Descartes simultaneously crystallized the commonsense view into his clear, sharply defined model of external, material nature surveyed by independent and totally dissimilar minds. The Galilean practice could not be abandoned because it alone could bring rapid discoveries; the Cartesian picture could not be abandoned because it alone enabled what was being done to be easily visualized. Accordingly a compromise was reached. The Cartesian world was the real world, but the way to study it was to 'abstract' various properties of its fundamentally integral objects— those properties which were capable of producing sensations in us— and study them separately. Thus science rendered formal homage to Descartes but all the time followed Galileo. Up to 1851 it could do so without embarrassment. In 1951 this is no longer possible; the later discoveries made in traversing the Galilean road have been found to be indescribable in Cartesian terms.

The Development of Psychology

The breakdown revealed itself most conspicuously at two independent points. The first in order of time was reached by psychology. In the real, external, extended Cartesian world there was no place for emotions, hallucinations, dreams, and such ephemeral things. Hence psychology was for long relegated to the study of the philosopher or at best restricted to what could be done in the laboratory of the physiologist from investigations of the human body. Everything that evaded these means of attack was dismissed as illusion. And from the Cartesian point of view this was reasonable enough. If you start with a postulated 'real' world which is the only worthy object of your analysis, it is sufficient to show that these mental vagaries do not belong to it to justify you in ignoring them. On the other hand, however, they are certainly experiences. Call a dream an illusion if you like; it still remains true that you have had it, that it was this and not that, and that if you are a John Bunyan, for example, it may have a profound and far-reaching effect on the future. And if an experience of motion can be successfully treated without regard at all to its material incarnation, why might not any other experiences, including those previously thought beneath consideration? With scientists unconsciously following the true Galilean prescription it was inevitable that sooner or later this step would be taken. Towards the close of the nineteenth century it actually was taken.

The immediately preceding chapters have described the modern scientific approach to psychology. Just as physical science makes use of concepts such as electrons, fields of force, space curvature, which are not at all evident in the material bodies our experience of whose behaviour they are designed to explain, so psychology makes use of such concepts as the unconscious, the censor, the super-ego, which are not at all evident even in the minds, let alone the bodies, our experience of whose behaviour they are designed to explain. Psychology as a science now stands on its own feet, drawing nothing from physics or physiology, except the example of how to proceed. The importance of this fact for our present consideration lies not at all in the validity or otherwise of the particular concepts now current. Those may very well, like electric fluids and phlogiston, pass into the limbo of discarded hypotheses as further knowledge is obtained. The significant thing is that psychology has clearly brought to light the fact that the primary data of science are not the constituents of a postulated external world but our actual experiences, some of which we formerly regarded as originating in that world but which now we must recognize as equally fit for scientific study whether it is possible to regard them as so originating or not. A modern Herschel can no longer look on a record of dreams as rubbish; it may be at least as important as a record of sunspots.

The Development of Physics

The second piercing of the nineteenth-century shell came from the growth of physics. In 1905 Einstein's first paper on the theory of relativity appeared, and that contained the idea which was to cause as irreparable a fracture as that produced by psychology. It may be described in terms of any of its numerous expressions; let us take the notion of simultaneity as an example. From the nineteenth-century point of view, to say that two events occurring in different places were simultaneous was to state a matter of objective fact. The events occurred in the real world, each occurred at a particular instant of universal time, and those instants were unambiguously either the same or not. All this was quite independent of the *observation* of the events. Whether anyone did or could experience them, and if so what his experience was, had nothing to do with their objective validity.

Nothing fatally at variance with this belief was known until the end of the nineteenth century, when it was seen to lead to a positive contradiction. One indubitable set of observations based on this idea showed in effect that two events were simultaneous; another indubitable set showed that they were not. The solution reached by Einstein depended on the shift of attention from the time of occurrence of an event to the means of observing that time. Instead of regarding the event and its time of occurrence as a primary datum and the process of determining the time as a means of ascertaining something independently established, Einstein took the operation of determining the time as yielding the primary datum and looked upon the phrase "time of the occurrence of the event" as merely a name for its result. The contradiction was then removed. The two sets of observations involved different operations and therefore could not be presupposed necessarily to give the same result. The business of physics was to take the results as given and to build up an idea of a world to which they were conformable; it was not to decide beforehand that there was a Cartesian world and then try to discover how it worked.

Now this in effect was identical with the transformation that had occurred in psychology. It was a return to the Galilean innovation of studying experiences and finding how one experience was related to another, instead of inventing an external world which 'existed' whether you could experience it or not. There is the difference, of course—very important for the practical task of advancing science but trivial for the understanding of it—that in psychology the experiences dealt with are in large part those that come spontaneously, without preparation or prompting on our part, whereas in physics we invariably experiment and measure, i.e. we set the conditions under which we observe and confine our experiences to the reading of manufactured scales. But essentially the procedure is the same. We start with experience and try to find out what sort of relations exist between experiences obtained under different conditions, instead of postulating a world independent of experience and then trying to discover how it is made up.

Psychology, by its very nature, is a far more difficult study than physics, and accordingly has made as yet no further advance relevant to our present concern. With physics, however, it is otherwise. Great progress has been made, and the situation is full of interest from the philosophical point of view. It is not easy to discard the habits of thought and the language of many generations, and, moreover, the majority of physicists are more anxious to push forward and enlarge the scope of their science than to interpret in the most fundamental terms the significance of what they are doing. Accordingly they often try to express the new knowledge in the old terms, with results that not unnaturally appear to the layman to touch absurdity. Let us look at the later developments of particle physics from this point of view.

Later Developments

We have seen that the old conception of the external world was that it was governed by inflexible laws that stood to observed phenomena in the relation of cause to effect. It was generally realized, however, that there were certain regularities in nature that could be explained (in the sense that events could be predicted from them, and trusted to occur) without any knowledge at all of the detailed mechanism by which they were produced. Thus, two gases, introduced independently into the same vessel, would, after a certain minimum lapse of time, be found each to fill the vessel uniformly, so that samples of the mixture taken from all parts of the vessel would be found to contain them in the same proportion. Now the molecules of the gas were conceived to be moving at random in all directions and undergoing frequent collisions. If we knew, at the moment of mixing, the position and motion of every molecule, then their positions at the end of the minimum time referred to would be calculable from the established laws of motion and they would in general be different for each set of initial motions. The molecules, however, could not be observed, so that we knew nothing of an essential factor in the determination of their later distribution: nevertheless we could predict with confidence, by an application of the principles of statistics, that uniform mixing would occur, without knowing anything at all about the whereabouts or the behaviour of one single molecule.

The curious conclusion was thus arrived at that in certain fields of study the results of all possible observations of certain kinds could be predicted with confidence, although the behaviour of the constituents of the real world, which perfectly obeyed the iron laws of cause and effect, remained quite inaccessible to knowledge. This position was reached three-quarters of a century ago.

If the Cartesian prejudice had not been so deeply ingrained in the minds of physicists they might have been led to suspect that since the 'real' world, with its causal laws, could hide itself so successfully without any diminution of their power of co-ordinating and predicting *experiences*, it might without sacrilege be deprived of its halo of 'reality' and recognized as a part of the conceptual machinery by which the observable results were calculated. Causal laws that caused only the unobservable might have been expected to lose something of their dignity when confronted with statistical laws that predicted the observable. But the time was not then ripe for such a revaluation. There was nothing contradictory in supposing that the unobservable was real and so physicists went ahead correlating experiences and paying lip-service to the world they were supposed to be discovering.

The developments of this century have strained this artifice to the breaking-point. First, the field of statistical relations has extended itself from the properties of gases over almost, if not quite, the whole scope of physics. *All* observable physical knowledge now appears to be statistical, so that the 'real' causal world lying behind it turns out to be completely unknown. Secondly, it is not only unknown but unknowable. By an analysis of the processes of observation that Einstein's discovery impelled physicists to undertake, it has been found that the means of observing the position of a molecule, or any other of the elementary constituents of the 'real' world, necessarily interferes with the means of observing its motion, so that an exact knowledge of these two properties, which is necessary to enable the future behaviour of the particle to be determined, can never in the nature of things be obtained. (This is known as *Heisenberg's Uncertainty Principle.*) Thirdly, the 'real' world is not only unknown and unknowable but inconceivable—that is to say, contradictory or absurd. For it turns out that when we apply our approved statistical methods to the particles in order to calculate what we can observe, there are some departments of physics in which those methods break down. We can make them effective only by ascribing to the particles properties not possessed by any imaginable objects at all. For instance, we must suppose that when, in the course of their motions, two particles change places, nothing has happened. It will not do to describe the second situation as exactly like the first; that leads to the wrong result. We must describe it as actually the same situation, and treat it statistically as one and not two possible arrangements of the particles.

Collapse of the Nineteenth-Century Philosophy of Science

The nineteenth-century 'real' world, then, has been forced, by the natural extension of scientific practice, from the foreground of our inquiry into the inaccessible background, where it has become unobservable and unpicturable. The causal laws once conceived to govern its behaviour, if there are such things, are unknowable and unnecessary for the purpose of expressing the correlations of our experience. Those correlations are expressed by statistical laws, in the application of which the supposed constituents of the 'real' world are endowed with whatever imaginable or unimaginable properties are necessary to give the right results. Such laws are in no sense commands or governing laws; they are simply descriptive statements of the regularities in our experience.

The Present Outlook

This leads to the following conclusion. If we regard the achievement of science as the discovery of relations between our experiences, we see a continuous and accelerated progress with no observable limitations. All scientific knowledge obtained up to the present is comprehended within this description, and all experience is potentially fit matter for scientific study. If, on the other hand, we try to maintain the old view that what science does is to discover the nature of an independent real world, we must add also that it has discovered that world to be unknowable and self-contradictory and its unveiling unnecessary for the purpose of understanding and predicting our experiences.

It is a striking indication of the tenacity with which traditional prejudices are retained that some physicists still express themselves in terms of the philosophy of 1851, accepting and acknowledging the absurdity of the world rather than giving up the belief that they themselves are a set of epiphenomena trying to discover its properties. Their number, however, is decreasing, and, except among adherents to materialistic economic doctrines that came to birth in the heyday of Victorian misconceptions and, being conceived to be necessarily infallible, are incapable of modification, the tendency is inevitably towards the wider and more actualistic (to avoid the misleading word, realistic) philosophy. The prospect could scarcely be viewed more optimistically than was that of a hundred years ago, but its horizon is immeasurably farther ahead.

THE END

INDEX

Hyphens between page numbers mean that references to the entry are made on all the pages covered. There may be more than one reference to an entry on a page.

317